The National Farm Survey, 1941–1943

State Surveillance and the Countryside in England and Wales in the Second World War

Surveillance of the English countryside depicted in original artwork: Hitler and Ribbentrop inspect an English scene through binoculars (source: PRO INF 3/1398, is Crown copyright and is reproduced with the permission of the Controller of Her Majesty's Stationery Office).

The National Farm Survey, 1941–1943

State Surveillance and the Countryside in England and Wales in the Second World War

Brian Short
School of Cultural and Community Studies
University of Sussex
UK

Charles Watkins
School of Geography
University of Nottingham
UK

William Foot
Defence of Britain Project
Council of British Archaeology
York
UK

and

Phil Kinsman
School of Geography
University of Nottingham
UK

CABI *Publishing*

CABI *Publishing* is a division of CAB *International*

CABI Publishing
CAB International
Wallingford
Oxon OX10 8DE
UK

Tel: +44 (0)1491 832111
Fax: +44 (0)1491 833508
Email: cabi@cabi.org

CABI Publishing
10 E 40th Street
Suite 3203
New York, NY 10016
USA

Tel: +1 212 481 7018
Fax: +1 212 686 7993
Email: cabi-nao@cabi.org

A catalogue record for this book is available from the British Library, London, UK

Library of Congress Cataloging-in-Publication Data
The National farm survey, 1941–3 : state surveillance and the
    countryside in England and Wales in the second World War / B. Short
    … [et al.].
        p.   cm.
    Includes bibliographical references.
    ISBN 0-85199-389-3 (alk. paper)
    1. Agriculture--England Statistics. 2. Agriculture--Wales
Statistics. 3. Agricultural surveys--England. 4. Agricultural
surveys--Wales. I. Short, Brian, 1944–   .
S455.N345   1999
630′.942--dc21   99-33180
                                                  CIP

ISBN 0 85199 389 3

Typeset in 10/12pt Garamond by Columns Design Ltd, Reading.
Printed and bound in the UK by Biddles Ltd, Guildford and King's Lynn.

Contents

Preface and Acknowledgements   vii

1 Introduction and Methodology   1
2 Agricultural, Social and Political Context of the Survey   15
3 The Organization of the National Farm Survey 1941–1943   41
4 Contemporary Critiques: Interpretation, Analysis and Policy   81
5 An Assessment of the National Farm Survey Data   113
6 The National Farm Survey Maps   143
7 Themes in the Wartime Rural Society and Economy of
        England and Wales   161
8 Conclusions   229

Appendix   239
References   245
Index   253

Preface and Acknowledgements

The genesis of this book lies in a chance meeting at a dinner held at the University College London in 1992 to celebrate the work and retirement of Hugh Prince. At that dinner Charles Watkins and Brian Short talked over their mutual interest in working on the National Farm Survey (NFS) records. Brian Short's interest had followed on from his earlier work on the Valuation Office 1910 material, and Charles Watkins' interest had arisen in the late 1970s when he had tried to access the National Farm Survey records to help reconstruct landed estate boundaries for his work on woodland management history. Soon afterwards they met to discuss a proposal with William Foot, then working at the Public Record Office (PRO), to obtain the necessary funding. An application to the ESRC to evaluate the potential of these records (what the agricultural historian G.E. Fussell called a 'New Domesday') was successful, and Dr Phil Kinsman was appointed to join William Foot (who had been seconded from the PRO, where he was then based) as Research Officers on the Project which was funded from 1994 to 1996.

We would like to thank colleagues in the Geography Subject Group and School of Cultural and Community Studies at the University of Sussex, and the School of Geography at the University of Nottingham, for their support throughout the work on which this book is based. The maps were prepared by Chris Lewis (Nottingham) and Susan Rowland and Hazel Lintott (Sussex). The authors would particularly like to thank Dr Sarah Tyacke, Keeper of the Public Record Office, and many of her staff, especially Alexandra Nicol, Geraldine Beech and Howard Davies. The work of the group was greatly facilitated by the PRO's encouragement and cooperation, its recognition of this as a 'PRO-friendly project', and the granting of editorial status to William Foot and Phil Kinsman.

We also wish to thank the following for their assistance at various stages of the project: Paul Barnwell (RCHME), Mrs J. Garrett (Ministry of Agriculture), Len Johnson, Professor Roger Kain, Dr John Martin, Professor Richard Munton, Rod Smith (Science Museum), and Martin Wingfield. We also acknowledge the assistance provided by all the staff at the County Record Offices and Libraries in England and Wales who responded to our questionnaire on their holdings of archival material relating to the NFS, and especially to the staff of the Buckinghamshire Record Office, East and West Sussex Record Offices, St Helens Local History and Archive Centre, the Hampshire Record Office, Professor Ted Collins and Dr Jonathan Brown of the Rural History Centre at the University of Reading, the Imperial War Museum, and the Library and Photographic Library of the Ministry of Agriculture. Many individuals also responded to our enquiries about the survey and provided help and assistance, including Paul and Hananja Brice and family, Professor Terry Coppock, the late Nigel Harvey, the late Laurie Lee, David Matless, Ann Nowill, Donald Sykes, the late Brian Wood and Mike Woods. And, as ever, our families and friends have provided encouragement and help throughout.

The following material, from the PRO, is Crown copyright and is reproduced with the permission of the Controller of Her Majesty's Stationery Office: frontispiece; Figs 1.1–1.5 inclusive; Fig. 3.6; Fig. 5.1; Figs 6.1 and 6.2; Figs 7.4–7.6 inclusive; Fig. 7.8; and Fig. 8.1. Figure 4.1 is copyright of the Science and Society Picture Library, Science Museum. Figure 7.2 is reproduced from S. Foreman, *Loaves and Fishes* (HMSO, 1989) by permission of the Ministry of Agriculture, Fisheries and Food. Figure 7.3 is reproduced by permission of the East Sussex Record Office, and Fig. 7.9 is reproduced by permission of Faber and Faber from R. Waller, *Prophet of the New Age* (1962).

Brian Short
Charles Watkins
William Foot
Phil Kinsman

July 1999

Introduction and Methodology   1

In the first half of the 20th century British agriculture was incorporated into modernity. This development was propelled by the growth of large-scale administration and bureaucratic systems of regulation that sought to move farming from an old order to a new. The requirement for surveillance over the nation's agriculture – felt for generations but now insistent in the face of the Great War and the following depression – became a central feature. Surveillance yielded information, an essential modern prerequisite for power and control. As political ideologies changed, so ideas about increasing intervention by the state blossomed within rural England as elsewhere. This in turn meant that an even greater requirement for modernist surveillance was posited for the smooth functioning of a state apparatus that could yield information to policy-makers, and which if possible was relatively cheap and widely acceptable.

Much of what follows in this volume is the outcome of that increased ideology of control over the nation's farming and food supply. Indeed, as well as food *production*, the requirement to control and ration food *supply and distribution* in wartime was equally pressing. Although this aspect of wartime control is not specifically dealt with in this volume we might note that a separate Ministry of Food was established in September, 1939, and that a three-volume official history of food was published after the war (Hammond, 1951, 1956, 1962).

The degree and manner of governmental intervention waxes and wanes according to changing economic, political and ideological conditions (Bowler, 1986: 124). But statistical surveys were a large part of the answer to governmental needs to control from a distance. Sir John

© CAB *International* 2000. *The National Farm Survey, 1941–1943*
(B. Short, C. Watkins, W. Foot and P. Kinsman)

Sinclair's *Statistical Account of Scotland* was begun in 1788 and the surveys of the Board of Agriculture (of which Sinclair was the first president in 1793) – surveys inspired and coordinated by Arthur Young – were an early version of this information gathering. As the 19th century unfolded, so the surveys of rural society and the rural economy grew in number and sophistication – sometimes organized directly by Parliament (as in the numerous Royal Commissions or parliamentary investigations, or statistical enquiries such as the 1873 'New Domesday' of landownership, or the 4 June Returns from 1866), sometimes by public societies (such as the Manchester Statistical Society in the 1830s, the Statistical Society founded in 1834, or as seen in the prize essays awarded by the Royal Agricultural Society). The capitalist state came increasingly to hold power through statistics – a word derived from the German *statistik*, or 'State-istics' (Shaw and Miles, 1979: 27). The late 19th and early 20th century work of G. Booth, A.L. Bowley and B.S. Rowntree, prompted by issues of poverty and inequality within the nation, stimulated governmental action towards the gradual evolution of the Welfare State and, importantly for the purposes of this volume, the incorporation of statistical enquiries into the government's own hands.

The movement towards statistical provision, surveys and analysis attracted statisticians from a wide range of backgrounds. Much of the early 20th century interest in statistics and surveys was fostered by mathematical statisticians who were connected intellectually with the Eugenics movement. F. Galton, K. Pearson and R.A. Fisher worked on statistical theory as an intrinsic element within British culture at this time, becoming interested in class differences, intelligence and heredity. Not all statisticians were working in this area: G. Yule and W.S. Gosset ('Student', the originator of the '*t*-test') are examples, with Gosset having earlier developed his *t*-distribution for working with small statistical samples whilst he was as an industrial scientist for Guinness Brewers.

By 1914 there were 12 institutes and two minor centres for agricultural research in Britain; by the 1920s, with agricultural productivity and large-scale state intervention in farming becoming issues, Fisher's work at Rothamsted research station had become well known (Russell, 1966: 272). Fisher was appointed to the newly created post of statistician in 1919 and some of his best-known work, including his *Statistical Methods for Research Workers* (1925), was carried out whilst in post at Rothamsted, before he left in 1933 to take up the Galton Chair of Eugenics at London University vacated by K. Pearson (MacKenzie, 1981: 210–213). He was succeeded by Frank Yates, and in the same year E.S. Pearson and others established an Industrial and Agricultural Research Section of the Royal Statistical Society. Such approaches to the study of agriculture were also fostered by the establishment of the Agricultural Research Council in 1931 (Cooke, 1981; Finney and Yates, 1981: 229).

Survey work had become widely recognized by the late 1930s. Dudley Stamp's Land Use Survey had been joined by Stapledon and Davies's grassland survey, completed just before the Second World War in the remarkably short time of 18 months (Waller, 1962: 234). Such attempts at comprehensive surveys were costly, however, and Yates and other statisticians were keen to make more use of the new sampling techniques. Sampling became more refined and in 1941 the Central Statistical Office was established, publishing its *Monthly Digest of Statistics* from 1946. A sophisticated shared framework of reference was now in place that embraced the concepts of social and economic investigation, large-scale surveys, and statistical analysis bent towards the practicalities of state control. A network had developed wherein the idea of a national survey of farming could be discussed and its bureaucratic as well as theoretical principles evaluated.

The Records of the National Farm Survey

In the year that the government's Central Statistical Office was opened, the National Farm Survey of England and Wales was begun. Carried out between 1941 and 1943, it was one of the fundamental components of the intervention by the state in the wartime economy and provides an excellent example of the surveillance held to be necessary in order to improve the productivity of agriculture at a time of potential food scarcity and crisis. As Lord Woolton (Minister of Food, 1940–1943) later stated: 'Those who found themselves in 1939 responsible for the conduct of the country in the rapidly changing conditions of war – when facts were more important than precedents – had very little to help them' (Moss, 1991: 3). The planning of the survey indeed began at a time of very real crisis, as the Battle of Britain was being fought out in the skies above the fields and farms, to be followed by the Blitz on London. It was launched in the weeks following the German invasion of Russia in the summer of 1941 and was being fully implemented by the time of the Japanese attack upon Pearl Harbor in December 1941, which brought America into the war. The survey ended at a time when the tide of war was turning fundamentally in favour of the Allied forces.

There were in fact two surveys of agriculture in these early years of the war. The earlier survey, which began in June 1940, will be described later in this volume, but without any great detail since, although it was important in its own right, it was in reality little more than a precursor to the greater National Farm Survey (NFS), which began the following year. By September 1940, during the first heavy air raids on London, Ministry of Agriculture officials were setting out proposals for the later survey, and a Farm Survey Committee was established in October, also concerned with planning the later survey. Circular 545 of 26 April 1941, then initiated the NFS, which ran over more than 2 years.[1] There is no real evidence that an official end was

FARM SURVEY

County_____ Code No._____
District _____ Parish _____
Name of holding_____ Name of farmer_____
Address of farmer_____
Number and edition of 6-inch Ordnance Survey Sheet containing farmstead_____

A. TENURE.

1. Is occupier tenant ... ...
        owner ... ...
2. If tenant, name and address of owner :—
_____
_____
_____

3. Is farmer full time farmer ... ...
        part time farmer ... ...
        spare time farmer ... ...
        hobby farmer ... ...
        other type ... ...
   Other occupation, if any :—

	Yes	No
4. Does farmer occupy other land ?		

Name of Holding	County	Parish

	Yes	No
5. Has farmer grazing rights over land not occupied by him ?		

If so, nature of such rights—
_____
_____

B. CONDITIONS OF FARM.

1. Proportion (%) of area on which soil is	Heavy	Medium	Light	Peaty

2. Is farm conveniently laid out ? Yes ... ...
        Moderately ...
        No ... ...

	Good	Fair	Bad
3. Proportion (%) of farm which is naturally			
4. Situation in regard to road ...			
5. Situation in regard to railway ...			
6. Condition of farmhouse			
Condition of buildings			
7. Condition of farm roads			
8. Condition of fences			
9. Condition of ditches			
10. General condition of field drainage			
11. Condition of cottages			

	No.
12. Number of cottages within farm area ...	
Number of cottages elsewhere	
13. Number of cottages let on service tenancy ...	

14. Is there infestation with :—	Yes	No
rabbits and moles		
rats and mice		
rooks and wood pigeons ...		
other birds		
insect pests		
15. Is there heavy infestation with weeds ?		

If so, kinds of weeds :—
_____
_____

	Yes	No
16. Are there derelict fields ?		
If so, acreage		

C. WATER AND ELECTRICITY.

Water supply :—	Pipe	Well	Roof	Stream	None
1. To farmhouse ...					
2. To farm buildings ...					
3. To fields					

	Yes	No
4. Is there a seasonal shortage of water ?...		
Electricity supply :—		
5. Public light		
Public power		
Private light		
Private power		
6. Is it used for household purposes ? ...		
Is it used for farm purposes ?		

D. MANAGEMENT.

1. Is farm classified as A, B or C ? ... ...

2. Reasons for B or C :—
      old age ... ...
      lack of capital ... ...
      personal failings ... ...

If personal failings, details :—
_____
_____
_____
_____
_____
_____
_____
_____
_____

	Good	Fair	Poor	Bad
3. Condition of arable land ...				
4. Condition of pasture ...				

	Adequate	To some extent	Not at all
5. Use of fertilisers on :—			
arable land ...			
grass land ...			

Field information recorded by
_____

Date of recording_____

This primary record completed by
_____

Date _____

FORM No. B496/E.I.   *15946. WL.46106/817. 8000 pads. 2/41. Wy.L.P. Gp.876.

Fig. 1.1. The Primary Farm Return of the National Farm Survey (NFS) (front page) (Form B496/E1). A blank copy (source: PRO MAF 38/256, is Crown copyright and is reproduced with the permission of the Controller of Her Majesty's Stationery Office).

ever brought to the Survey, but by July 1943, thoughts were turning to the analysis of the collected data by the Advisory Economists attached to the project, although the final summary report did not appear until August 1946.

The methodology and the context of the compilation of these records will be described in some detail in this volume. The whole NFS was devised as a decentralized operation, to be conducted under the auspices of the County War Agricultural Executive Committees (CWAECs) in each locality. Following initial preparations, visits to each farm were undertaken by CWAEC District Committee members, the 'maids-of-all-work' at this time (Rutherford and Bateson, 1946). The outcome of such visits was the completion of the Primary Farm Return or Record (the terms are interchangeable), which provided detailed information on tenure, the conditions on the farm, water and electricity supplies, and farm management (Fig. 1.1).[2]

The annual 4 June Returns were completed as usual during the war but for 1941 two supplementary enquiries were added to the normal questions concerning crops and livestock. One was to ascertain more detail concerning small fruit and horticultural produce (referred to herein as the Horticultural Return) and the other was the 'Supplementary Form' detailing motive power on the holding, rents and length of occupation. When all these had been returned to the Ministry as usual by the farmers, via the Ministry's teams of Crop Reporters, the Ministry then made available from its wartime statistical branch based at Lytham St Anne's to the CWAECs all 4 June 1941 Census Returns showing crops and grass, labour, livestock and horses. The confidentiality accorded to these returns since their inception in 1866, and for all years since 1941 to the present, was suspended in the interests of the wartime emergency. District Committee members, who were often near-neighbours of the farmers concerned, could now see quite plainly what was being produced around them in their neighbourhoods (Figs. 1.2–1.4).[3] For the first time, too, a measure of the completeness of the 4 June Returns could be supplied, and evidence confirming that there were many farms avoiding the census was now forthcoming. The survey of Buckinghamshire in 1938 had found 349 occupiers of land over 1 acre whose names were not on the Ministry's list and altogether over 13,000 acres were omitted (Thomas and Elms, 1938: 11; Coppock, 1955: 14).

The final component of the survey was the series of large-scale maps on which were inscribed details of farm boundaries. These were either OS 1:2500 sheets photographically reduced to 1:5000, or were 1:10,560 sheets. Their value in 1941 was enhanced by the frequent marginalia added to the map margins, which detailed occupiers of the land shown on the sheets. Either by a colour wash or by colouring in the actual farm boundaries, the farm structures could now be seen (Fig. 1.5).[4]

The records of the first (1940) Survey are now available only as county summaries and detailed farm records do not appear to have survived.[5] The 1941–1943 Survey records were recalled from the Provincial Advisory

MINISTRY OF AGRICULTURE AND FISHERIES
THE DEFENCE REGULATIONS, 1939, AND THE AGRICULTURAL RETURNS ORDER, 1939.
RETURN WITH RESPECT TO AGRICULTURAL LAND ON 4th JUNE, 1941.

Fill in both Forms 398/S.S. (Green) and C 47/S.S.Y. (Blue).

The name of the holding, if not included in the postal address, is shown here ➤

Your Acreage last year was Crops & Grass / Rough Grazings

CROPS AND GRASS	Statute Acres
1	Wheat	
2	Barley	
3	Oats	
4	Mixed Corn, with Wheat in mixture	
5	Mixed Corn, without Wheat in mixture	
6	Rye	
7	Beans, winter or spring, for stock feeding	
8	Peas, for stock feeding, not for human consumption	
9	Potatoes, first earlies	
10	Potatoes, main crop and second earlies	
11	Turnips and Swedes for fodder	
12	Mangolds	
13	Sugar Beet	
14	Kale, for fodder	
15	Rape (or Cole)	
16	Cabbage, Savoys, and Kohl Rabi, for fodder	
17	Vetches or Tares	
18	Lucerne	
19	Mustard, for seed	
20	Mustard, for fodder or ploughing in	
21	Flax, for fibre or linseed	
22	Hops, Statute Acres, not Hop Acres	
23	Orchards, with crops, fallow, or grass below the trees	
24	Orchards, with small fruit below the trees. (See also next page)	
25	Small Fruit, not under orchard trees. (See also next page)	
26	Vegetables for human consumption (excluding Potatoes), Flowers and Crops under Glass (See also next page)	
27	All Other Crops not specified elsewhere on this return or grown on patches of less than ¼ acre	
28	Bare Fallow	
29	Clover, Sainfoin, and Temporary Grasses for Mowing this season	
30	Clover, Sainfoin, and Temporary Grasses for Grazing (not for Mowing this season)	
31	Permanent Grass for Mowing this season	
32	Permanent Grass for Grazing (not for Mowing this season), but excluding rough grazings	
33	TOTAL OF ABOVE ITEMS, 1 to 32 (Total acreage of Crops and Grass, excluding Rough Grazings)	
34	Rough Grazings—Mountain, Heath, Moor, or Down Land, or other rough land used for grazing on which the occupier has the sole grazing rights	

	LIVE STOCK on holding on 4th June, including any sent for sale on that or previous day	Number (in figures)
35	Cows and Heifers in milk	
36	Cows in Calf, but not in milk	
37	Heifers in Calf, with first Calf	
38	Bulls being used for service	
39	Bulls (including Bull Calves) being reared for service	
40	OTHER CATTLE — 2 years old and above — Male	
41	2 years old and above — Female	
42	1 year old and under 2 — Male	
43	1 year old and under 2 — Female	
44	Under 1 year old:— (a) For rearing (excluding Bull Calves being reared for service)	
45	(b) Intended for slaughter as Calves	
46	TOTAL CATTLE and CALVES	
47	Steers and Heifers over 1 year old being fattened for slaughter before 30th November, 1941	
48	SHEEP OVER 1 YEAR OLD — Ewes kept for further breeding (excluding two-tooth Ewes)	
49	Rams kept for service	
50	Two-tooth Ewes (Shearling Ewes or Gimmers) to be put to the ram in 1941	
51	Other Sheep over 1 year old	
52	SHEEP UNDER 1 YEAR OLD — Ewe Lambs to be put to the ram in 1941	
53	Ram Lambs for service in 1941	
54	Other Sheep and Lambs under 1 year old	
55	TOTAL SHEEP and LAMBS	
56	Sows in Pig	
57	Gilts in Pig	
58	Other Sows kept for breeding	
59	Barren Sows for fattening	
60	Boars being used for service	
61	ALL OTHER PIGS (not entered above) — Over 5 months old	
62	2—5 months	
63	Under 2 months	
64	TOTAL PIGS	
65	POULTRY If none, write "None" — Fowls over 6 months old	
66	Fowls under 6 months old	
67	Ducks of all ages	
68	Geese of all ages	
69	Turkeys over 6 months old	
70	Turkeys under 6 months old	
71	TOTAL POULTRY	
72	GOATS OF ALL AGES	

LABOUR actually employed on holding on 4th June. The occupier, his wife, or domestic servants should not be entered.

73	WHOLETIME REGULAR WORKERS If none, write "None" — Males, 21 years old and over	
74	Males, 18 to 21 years old	
75	Males, under 18 years old	
76	Women and Girls	
77	CASUAL (SEASONAL or PART-TIME) WORKERS — Males, 21 years old and over	
78	Males, under 21 years old	
79	Women and Girls	
80	TOTAL WORKERS	

Form No. 398/S.S.   M.14000 6/41 (32-6831)

Fig. 1.2. The 4 June 1941 Return (page 1) (Form 398/SS). A blank copy (source: PRO MAF 38/256, is Crown copyright and is reproduced with the permission of the Controller of Her Majesty's Stationery Office).

SMALL FRUIT (See Instruction 2 (c))	Statute Acres
81 Strawberries	
82 Raspberries	
83 Currants, black	
84 Currants, red and white	
85 Gooseberries	
86 Loganberries and Cultivated Blackberries	
87 Total Acreage of Small Fruit (This total should equal the total of Nos. 24 and 25)	

VEGETABLES FOR HUMAN CONSUMPTION. FLOWERS. CROPS UNDER GLASS.
(See Instruction 2 (d))

		Statute Acres
88	Brussels Sprouts	
89	Cabbage, Savoys, Kale, and Sprouting Broccoli	
90	Cauliflower or Broccoli (Heading)	
91	Carrots	
92	Parsnips	
93	Turnips and Swedes (not for fodder)	
94	Beetroot	
95	Onions	
96	Beans, Broad	
97	Beans, Runner and French	
98	Peas, Green, for Market	
99	Peas, Green, for Canning	
100	Peas, Harvested dry	
101	Asparagus	
102	Celery	
103	Lettuce	
104	Rhubarb	
105	Tomatoes, growing in the open	
106	Tomatoes, growing in GLASSHOUSES	
107	Other Food Crops growing in GLASSHOUSES	
108	Flower Crops growing in GLASSHOUSES	
109	Crops growing in FRAMES (fruit, vegetables, flowers, and plants).	
110	Hardy Nursery Stock	
111	Daffodils and Narcissi, not under glass	
112	Tulips, not under glass	
113	Other Bulb Flowers, not under glass	
114	Other Flowers, not under glass	
115	TOTAL (this total should equal No. 26)	

Name of Farm or Farms to which this return relates :—

(Note.—Any person occupying more than one holding should make a separate return for each holding which is farmed separately. Holdings farmed together, or outlying pieces of land farmed with the main holding, should be included in one return.)

HORSES on holding on 4th June	Number (in figures)
116 Horses used for Agricultural Purposes (including Mares kept for breeding) or by Market Gardeners (a) mares	
117 (b) geldings	
118 Unbroken Horses of 1 year old and above (a) mares	
119 (b) geldings	
120 Light Horses under 1 year old	
121 Heavy Horses under 1 year old	
122 Stallions being used for service in 1941	
123 All Other Horses (not entered above)	
124 TOTAL HORSES	

STOCKS OF HAY AND STRAW ON HOLDING ON 4th JUNE

		Tons
125	Hay	
126	Straw	

CHANGES OF OCCUPATION

Give particulars of any land in your occupation on 4th June, 1940, but since given up and therefore excluded from Items 33 and 34 on page 1.

	To whom given up:— Name and Address of New Occupier (if known)	Arable Land and Permanent Pasture	Rough Grazings
		acres	acres
127			

Give particulars of any land taken over since 4th June, 1940, and now in your occupation and therefore included in Items 33 and 34 on page 1.

	Name and Address of Former Occupier	Arable Land and Permanent Pasture	Rough Grazings
		acres	acres
128			

*Signature of Occupier

Postal Address :—

Date

* If signed by an Accredited Agent the name of the actual occupier must also be shown.

Fig. 1.3. The Horticultural Return incorporated in the 4 June 1941 Return (Form C51/SSY). The blank copy shown here comprises page 2 of the Return (source: PRO MAF 38/256, is Crown copyright and is reproduced with the permission of the Controller of Her Majesty's Stationery Office).

S.F.

MINISTRY OF AGRICULTURE AND FISHERIES

AGRICULTURAL RETURN, 4th JUNE, 1941.

LABOUR ON 4th JUNE (Supplementary Questions).

			Number
	Of the REGULAR workers returned on page 1 (Questions 73—76) how many are:—		
129	WHOLE TIME {father, mother, son, daughter, brother, sister of}	male	
130	FAMILY WORKERS {occupier or his wife, but not other relations}	female	
	Of the CASUAL workers returned on page 1 (Questions 77—79) how many are:—		
131	EMPLOYED ON THE HOLDING THROUGHOUT THE }	male	
132	YEAR BUT FOR ONLY PART OF THEIR TIME }	female	

MOTIVE POWER ON HOLDING ON 4th JUNE.

	FIXED OR PORTABLE ENGINES (Excluding Motor Tractors)	Number in figures	Horse Power of each
133	Water Wheels or Turbines in present use		
134	Water Wheels not in use, but easily repairable		
135	Steam Engines		
136	Gas Engines		
137	Oil or Petrol Engines		
138	Electric Motors		
139	Others (state kinds)		

	TRACTORS	Number in figures	Horse Power of each	Make or Model of Tractor
140	Wheel Tractors for field work			
141	Wheel Tractors for stationary work only			
142	Track laying Tractors			

NOTE.—Subject to the special Question No. 134 engines or tractors that have been discarded or worn out should not be included.

RENT

ANNUAL RENT PAYABLE FOR THE HOLDING TO WHICH THIS RETURN RELATES.

			£
143	State the actual rent payable during the current year (i.e., the contract rent less any abatements but including any interest payable on improvements).		
144	If the holding is owned by you, give the best estimate you can of the annual rental value		
		Acres	£
145	If the holding is partly owned and partly rented by you, state:— Acreage of land which you own and its estimated rental value and		
146	Acreage of land which you hold as tenant and the rent payable (for definition of rent see Question No. 143)		

LENGTH OF OCCUPATION OF HOLDING.

		Years
147	How many years have you been the occupier of the holding to which this Return relates ?	
	or	
148	If you have occupied parts of the holding for different periods, give length of occupation for each { Part 1..................acres..................years Part 2..................acres..................years Part 3..................acres..................years	

FOR OFFICIAL USE ONLY.

Fig. 1.4. The Supplementary Form incorporated in the 4 June 1941 Return (Form 398/SS). The blank copy shown here comprises page 3 of the Return (source: PRO MAF 38/256, is Crown copyright and is reproduced with the permission of the Controller of Her Majesty's Stationery Office).

Fig. 1.5. An example of an annotated National Farm Survey (NFS) Ordnance Survey (OS) sheet (extract). Hartfield, East Sussex, on the northern boundary of Ashdown Forest. Cotchford Farm on the east of the map was the home of A.A. Milne, and Poohsticks Bridge lies between Cotchford and Posingford Woods. The OS reference numbers as well as the acreages have been pencilled in (source: PRO MAF 73/41/16, is Crown copyright and is reproduced with the permission of the Controller of Her Majesty's Stationery Office).

Centres, where they had resided since the war, and were assembled by Ministry archivists at Hayes in the late 1950s as five related, but separately packaged, groups of records. The forms were then sent to the Public Record Office (PRO) at Chancery Lane in 1959, and the maps in 1970, and transferred to the new Kew office in 1976–1977. At the PRO there are now the four different sets of forms and a set of maps – and thus, theoretically, five different elements yielding data for each holding in England and Wales of 5 acres or more. Because the Ministry had been careful to guard the confidentiality of the material – especially the grades relating to the management of the farms – the records remained closed to public inspection for a 50-year period, deemed to have finished in January 1992.[6] At that date an unparalleled cross-sectional view of British farming in wartime became available. Not only was there a survey that had been carried out by knowledgeable personnel for each farm, there was also a detailed farm boundary map, and for the year 1941 modern scholars also have available the full 4 June Returns, rather than the more familiar parish abstracts.

In order to evaluate these records, an application was made to the Economic and Social Research Council in 1993 and an award was made for the period October 1994 to March 1996.[7] The remainder of this first chapter will elaborate on the methodologies of that evaluation and set the scene for the structure of the volume as a whole.

Research Methodologies

The previous section outlined the diversity of records that together form the NFS archive at the PRO. At the outset of our study, decisions had to be taken on the sampling methodology to approach the enormous data set, in order to assess the quality and usability of the material for research purposes. An initial decision was taken to transcribe approximately 1% of the records comprising the NFS, and to compile this sample into three databases consisting of two regional samples and one national sample. In total this entailed transcribing all the available data from the NFS for about 3000 holdings in England and Wales.

The two regional samples consisted of groupings of parishes that were chosen as being based on the previous research experience of the authors, allowing some general background knowledge of the agricultural and social histories of the areas to be related to the NFS material. One regional sample was based in south-east England, in East and West Sussex, and this is referred to throughout this volume as the *Sussex Sample*. This sample consisted of 1200 holdings covering large areas of the South Downs and parts of the Sussex Weald. The other regional sample, referred to throughout as the *Midlands Sample*, consists of 480 holdings in three groups of parishes: one in Nottinghamshire, one in the Peak District of Derbyshire, and one in Herefordshire. This tripartite division allowed an examination of the NFS in parishes heavily influenced by neighbouring built-up areas (Nottinghamshire), characterized by upland farming systems (Peak District) and a remote mixed farming area (Herefordshire) (Fig. 1.6).

For the *National Sample* one parish was selected from every county in England and Wales, and the location and name of the chosen parish is shown in Fig. 1.6. This sample allowed the exploration of variation in the quality of NFS data across the country as a whole, as well as the investigation of differences in the collection and quality of data between the 11 different Agricultural Advisory Provinces at that date. The parishes were selected to provide a wide range of farming and landscape types. The National Sample is the largest of the three samples, with 1450 holdings. Overall, 3130 holdings were included in the three samples and it is the thorough examination of the NFS data for these holdings that forms the empirical basis for this volume. The choice of samples has therefore allowed the study of the survey material, its collection, the relationships between different elements of the NFS and its usability in regional contexts, as well as the variations in the quality and nature of the data across England and Wales.

The NFS data were transcribed on to a database, using laptop computers at the PRO. The goodwill and active assistance of staff at the PRO was crucial in allowing the required access to large quantities of documents and maps, and the editorial status granted to members of the project was invaluable. The laptops were essential as the documents were only accessible within the PRO reading rooms, where personal computers could not be used. Microsoft Access 2.0 for Windows was then chosen as the principal software for data collection and analysis. Access 2.0 is a relational database management system that facilitates the storage of data in a number of tables linked by particular fields.

The database constructed to transcribe and analyse the NFS data comprised 16 tables that generally followed the structure of the forms located in PRO MAF 32, although details on farm labour were gathered from two separate NFS forms and placed in one table. An additional table was constructed to hold data from the maps in PRO MAF 73. Unlike the other tables, this latter did not consist of transcribed data, but rather contained variables designed to allow for an assessment of the nature and quality of the maps. All the tables in the database were linked together by a single farm reference code. The nature of the responses sought by the original framers of the survey was used to structure the tables, rather than the responses actually given. The tables had to be designed with sufficient flexibility to accommodate the actual responses given while maintaining a balance with enough consistency to allow meaningful analysis. This was especially important, since both farmers and farm surveyors devised a wide variety of subversions and misunderstandings of the forms and how to complete them. The aim of the database was to transcribe the data in order to evaluate its quality, and

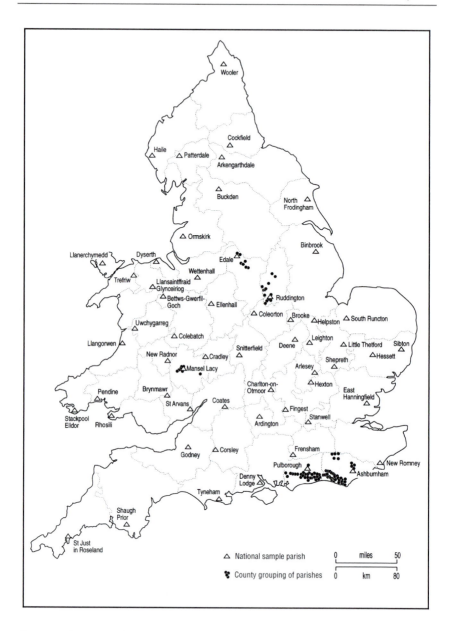

Fig. 1.6. The National and Regional Sample Parishes.

the data consist almost entirely of what appears on the forms, although some obvious and glaring errors were corrected (and a note made of such corrections).

The transcription of information from the four NFS forms outlined above on to the database could take between 7 and 35 min, depending on the amount of data present on the original forms, the complexity of the holding's relationship to other holdings, and the clarity of the information on the forms. An estimate of the average amount of time taken per holding was about 15 min, and the number of records transcribed during a working day varied between 12 and 25, with an average of about 16 holdings per day.

The Structure of this Volume

The aim of this volume is to explore the complexity of the NFS data, to demonstrate the construction of the survey data, and to examine the worth of the NFS as a source for the study of rural England and Wales in the mid-20th century. Chapter 2 considers the agricultural, social and political context of the survey, concentrating in particular on changing agricultural policies, the development of organizational structures and the rise of a survey mentality prior to the war. Chapters 3 and 4 examine in detail the organization of the NFS during the war, the perception and reception of the NFS by contemporaries, and its publication and subsequent archival treatment. Chapter 5 assesses the quality of the NFS data on each of the four constituent forms; each variable is discussed in turn and in detail, and problems of interpretation are analysed. In Chapter 6 attention turns to the analysis of the NFS maps and consideration of their value as a source at local, regional and national scales. Chapter 7 examines the potential value of the NFS for the study of six selected themes: land ownership; farm size and structure; agricultural mechanization; women and agriculture; the military use of land; and the plough-up campaign. These contrasting themes have been selected primarily to demonstrate the problems and the benefits of using this rich source of wartime data.

Notes

1 PRO MAF 38/210 (10). An early draft of the Circular is in PRO 38/207 (6).
2 Referred to officially as Form B496/E1, and retained in the PRO under Class MAF 32. See Chapter 4 for a full assessment of these forms and all other materials arising from the Survey.
3 Form C47/SSY p.1, retained in the PRO also as MAF 32, as distinct from the abstracts for parishes in MAF 68.
4 The maps are to be consulted as PRO MAF 73. Chapter 5 offers an analysis of the quality of these maps. See also W. Foot, *Maps for Family History* (PRO 1994), 52–66.
5 The county summaries can be seen in PRO MAF 38/213.
6 For more information on the issues surrounding the confidentiality of these records, see Chapter 3.

7 ESRC reference number R000235259. Dr P. Kinsman was employed as Research Fellow on the grant and based at the University of Nottingham, and Mr A.W. Foot was seconded from the PRO for the duration of the project.

Agricultural, Social and Political Context of the Survey   2

A key theme in the years leading up to the great agricultural changes of the Second World War was the translation of policies that had been framed at the national level into local action. Hence much of the following account of the context of the survey must interlace the national (and indeed international) policy scene and the wartime implementation of such policies at the scale of the rural locality. A review of the interwar policy decisions of the Ministry of Agriculture and the structure of the Ministry itself is thus followed by an account of three specific policy issues: the growing interest in rural surveys; the installation of the County War Agricultural Executive Committees (CWAECs); and the plough-up campaign. The chapter ends by touching on more general wartime farming policies.

Interwar Policy Developments

Following the introduction of price support and the intensification of agriculture during the First World War, a Royal Commission was established in July 1919 to consider the future of cereal prices. At the end of 1919 three reports were submitted: one signed by 12 members, one signed by 11 members and one signed by a lone member. The majority report recommended the further continuation of guaranteed minimum prices with a 4-year period of notice to end the guarantees. The minority report felt that guaranteed prices were now unnecessary. In the Agriculture Bill of June 1920 the majority report was incorporated and deficiency payments for wheat and oats were to be maintained. This was confirmed in the Agriculture Act of

December 1920, but the resumption of massive imports from Canada, the Argentine, Australia and India brought wheat prices down from 89s 3d per quarter in June 1921 to 45s 8d in December. With such vast price reductions it became clear to the embarrassed government that a huge liability of expense had been entered into which could not now be afforded. In June 1921 the Minister of Agriculture, Sir Arthur Griffith-Boscawen, announced the immediate repeal of those sections of the 1920 Act that had guaranteed the minimum prices, thereby instigating 'the Great Betrayal' which left its mark in creating the farming community's suspicion of future government actions. The early 1920s became instead a period of belief in agricultural prices determined by market forces (Whetham, 1974, 1978: 109–141; Penning-Rowsell, 1997: 176–194).

Rather than subsidize arable crops, Baldwin's government advised farmers to grow less corn and move more strongly into the production of meat and milk. There were some minor moves, such as the subsidy to establish sugar beet, which resulted in the expansion of that industry amongst what A.G. Street called 'the spoilt babies of British farming' in East Anglia during the later 1920s (Whetham, 1978: 165–169; Douet, 1996). Then came the Depression of 1929–1933. Downward pressures on corn prices were followed by reductions in farm incomes generally as the agricultural prices index reached its low point in the winter of 1932/33, although the calamitous times for cereal growers (especially on the heavy clays) were in many senses a boon to those who fed livestock.

With the defeat of the Conservative government in May 1929, and particularly under the National government headed by Ramsay McDonald, state intervention began in earnest. Agricultural Marketing Acts were passed in 1931 and 1933, and Walter Elliot took on ministerial responsibility for agriculture in September 1932 with a firm belief in national coordination. The Wheat Act 1932 provided a deficiency payment scheme (the first in peacetime) which stabilized the acreage of wheat in the following years; subsidies were rendered permanent for sugar beet and fat cattle; and marketing schemes were introduced for potatoes (1934), bacon pigs (1934) and milk (1933) to join that for hops which had been introduced in 1932 with a Hops Marketing Board. As well as these support and marketing measures, limited import duties were introduced. The full panoply of State involvement was now being revealed, and the Milk Marketing Board in particular has been seen as a good example of emerging corporatism at this time (Winter, 1996: 90–94).

In the later 1930s farm incomes began to rise again, in part related to the general recovery in world prices. Nevertheless, by 1939 British agriculture was deemed moribund after years of landed depression and subsequent lack of investment. Traditional methods, impoverished farms and derelict fields and buildings were the norm, and over much of England and Wales the countryside resembled a rabbit-infested wilderness rather than a productive farming scene. Furthermore, collective organization by the farming community itself – the panacea for progress adopted by the National Farmers' Union (NFU) rather than state paternalism – had not shown any marked development (Cooper, 1989: 57, 78–80). Yields had barely changed over the last 50 years.

In 1939 Earl De La Warr, Parliamentary Secretary to the Minister of Agriculture 1930–1935 and by then President of the Board of Education, put forward a memorandum on agricultural policy. The emphasis of his message was on the raft of drastic measures required for action against the deterioration of land and the decline of the agricultural population, and its spirit mirrored the realization that had been growing during the 1930s that only fast and effective intervention by the state in the farming industry could effectively introduce change. Within the next 5 years, he explained, the country would need a concerted programme aimed at increasing the quality and quantity of milk production, the improvement of grassland, improvements in slaughtering of home-killed meat, the reorganization of fruit and vegetable marketing, modifications to the agricultural credit scheme, re-equipment grants, the purchase and re-equipment of neglected or derelict land by the State, and the improvement of rural housing. The overall emphasis, however, was on milk and grassland, and on the rescue from years of depression of all sectors of the farming community:

> If there were, or could be, an enquiry into the amount of farm stock owned by the auctioneers, and of land in the hands of the banks, the result would probably be a surprise and a shock to us all – even more so if we could add the number of farmers who are so in debt to the merchants that they have virtually to pay what they are asked for feeding stuffs.[1]

De La Warr himself was an advocate of agricultural progress. He had inherited his Sussex estate in 1915, had begun to make silage in 1920, had replaced his Shorthorn herd with Jerseys and, following a meeting with George Stapledon in 1933, he had introduced ley farming. He had entered into a coterie of agricultural enthusiasts who debated through the 1930s and 1940s the most appropriate methods of dealing with the soil. Here he mixed not only with Stapledon, but also with such men as Viscount Lymington, Lord Northbourne and H. Rolf Gardiner (1902–1971) at Springhead, Dorset. Radical ideas were traded against the more established views which inclined towards larger-scale capital intensification rather than the 'cycle of life' ideas of Lymington, which were based on a partial and early understanding of ecological principles, elided with right-wing views on English nationalism (Matless, 1998: 103–172).

However, such debates among the landed intelligentsia had largely been cut short from about 1937. Writings concerned with improving nutritional standards and helping the agricultural industry in line with such help proffered to other industrial sectors, and which had aimed to halt rural depopulation and land dereliction, changed instead to a discourse on national defence. In 1926 Stanley Baldwin's government had announced its decision that there was no case on defence grounds to justify the expenditure necessary to induce farmers in peacetime to produce 'more than economic considerations dictated' (Murray, 1955: 47).[2] But as the expected hostilities within Europe gathered force, preparations for increasing food production by whatever methods and from whatever available resources were put into action. During the 1930s there had certainly been debates about the need to produce more home-grown food, and such debates often mingled uneasily with those provoked by the 'Humus school of thought' of Rolf Gardiner, Viscount Lymington or Lord Northbourne (Gardiner, 1943). In 1935, for example, George Stapledon (1935: 14) had written:

> Last time we were caught out completely – a very small proportion of our land area was in fertile condition, or suitable for crop production; few resources, limited knowledge, considerable dissipation of energy, large numbers of farmers ignorant of the art of cultivation – we had only just begun to get into our stride when mercifully our endeavours became no longer necessary.

Few could argue with the substance of Stapledon's comments. In 1935 Walter Elliot, the Minister of Agriculture, appointed a committee to consider the question of food production in time of war, in response to the announcement of the existence of a German airforce and Hitler's reintroduction of conscription. By 1936 a Food (Defence Plans) Department had been established within the Board of Trade, and in 1937 Morrison, now Minister of Agriculture, established an interdepartmental committee 'to make definite proposals for increasing the productivity of our own soil, with a view to increasing food production in time of war' (Murray, 1955: 53). The committee's recommendations, embodied in the 1937 Agriculture Act, included increased advice on soil fertility, grants to encourage the use of lime and slag, encouragement for the care of minor watercourses and ditches, a campaign to eradicate animal diseases, and an improvement in the guaranteed price system for wheat.

Overall by 1939 farmers had seen several Agricultural Marketing Acts passed and 17 boards or associations of producers were now in being. Imports were being regulated, and the use of subsidies, price insurances and deficiency payments was becoming acceptable. A 'policy of control' rather than a 'policy of cheapness' had been installed (Astor and Murray, 1933: 30). This effectively meant that state intervention in peacetime was a well-established concept by 1939. Looking back over the previous years, Venn (1939: 49)[3] opined that the changes had come about:

> this time not despite a policy of laissez-faire, but as a result of considered actions and preferential treatment of an all-embracing character meted out by the State with the approval of the majority of the nation.

Not everyone would have agreed that there was a considered policy. Astor and Rowntree in 1938 saw rather a succession of *ad hoc* measures imposed ever since the Armistice to avert ruin and to help particular sectors of the farming community. Indeed, the distinction has rightly been drawn between intervention and planning – much of the early intervention in the 1930s was concerned with particular markets rather than with an overall strategy, with the result best characterized as 'quasi-corporatism' (Astor and Rowntree, 1938; Cox *et al.*, 1985; Cooper, 1989: 143). Hammond (1954: 7) noted that:

> These measures of assistance, though each and all can be defended on grounds of expediency, did not add up to anything that could be called a *policy* for British agriculture. They rested on no coherent appraisal of the role that its various branches ought to play in the national economy, whether in peace or war.

Despite the changes during the earlier part of the decade, by 1938 the UK was still dependent on overseas sources for no less than 70% by value of its food supplies, and some 23 million tons of human and animal food and fertilizers were imported annually. When the anticipated interruptions to sources of continental food were taken alongside shortages of foreign currency and the carrying capacity of the remaining shipping, the issue assumed prime importance. Other issues loomed large too. Could the half a million farmers be sufficiently well organized to receive and exploit supplies of labour, equipment, fertilizers, etc.? How best could they be motivated, and if financial incentives were needed, was this best applied to price increases or input subsidies? And were the costs that were anticipated in such a massive intervention going to yield returns that were sufficiently acceptable, given that scarce factors of production in wartime might be deployed elsewhere?

The Organizational Structure of the Ministry of Agriculture

The relations between agriculture and the State, which, as H.E. Dale remarked in 1939, are 'a wide field over which armies of writers and speakers have marched, leaving it a little muddy', are mediated by the Civil Service (Dale, 1939: 2). In 1889 the Board of Agriculture had been created with a president with Cabinet status. In 1903 fisheries were added, and in 1919 the Ministry of Agriculture and Fisheries (MAF) was fully constituted. Dale wrote also of the relatively low place in the Civil Service hierarchy that MAF officials enjoyed compared with the Home Office or Colonial Office, since they were appointed for their technical qualifications rather than their Oxbridge educations. This changed rapidly but briefly with the great thrust of State intervention during the First World War, but in the 1920s the Ministry remained comparatively small and static, until the interwar development of

State control ensured that the numbers of civil servants multiplied rapidly. In 1927 the Ministry was organized in seven divisions and employed 1400 people; by 1937 there were 11 divisions and 1600 employees, despite the loss of a large block of work on tithes. There were also now more university-trained members in its administrative class. Dale (1939: 13) thought this an excellent development, since:

> If a Principal Assistant Secretary in the Ministry of Health wants Treasury agreement to some slightly controversial proposal, he is likely to settle it more quickly and easily if his 'opposite number' at the Treasury is a man whom he knew when they were both undergraduates at Oxford and has known familiarly ever since then. They talk the same language, and lay their minds alongside each other with no risk of misunderstanding.

During the interwar years the Ministry had developed responsibilities for research and education, land settlement, land drainage, pest and disease control, animal health and improvement, agricultural labour and wages, marketing, commodity subsidies and pricing schemes. In discharging such responsibilities, links with organizations such as the NFU, the marketing boards and the Central Landowners Association had been forged which were now to be extremely important in the war. Between 1912/13 and 1944/45 gross expenditure by the Board or Ministry rose from £387,000 to £61.5 million, of which £56.6 million was related to the wartime emergency (Murray, 1955: 320–323).

By the outbreak of war, it seemed very clear that the Ministry had various central functions to perform, but also that regional decentralization of administration related to the complex agricultural community was vital. As Reginald Dorman-Smith said, 'There will be no farming from Whitehall this time' (Martin, 1992: 78). The central functions, relayed from Whitehall but soon from Blackpool or Bournemouth (whence most of the staff were sent for the duration of the war), included: the formulation of general policies and objectives in association with other departments, and ensuring that the appropriate legal machinery was in place; the broad allocation of production targets and supplies across the country; publicity and propaganda; and the organization and coordination of the 61 local CWAECs. In all, about 1900 circulars went from the Ministry of Agriculture to the CWAECs. The Executive Officer attached to each CWAEC was not a civil servant but one of the Ministry's Land Commissioners attached to each county committee to advise on the exercise of the legal powers delegated by the Minister, especially on the issue of compulsory orders and eviction (Winnifreth, 1962: 29).

A Chief Agricultural Adviser and Liaison Officer was appointed to headquarters on the outbreak of war, and under him 12 Liaison Officers were appointed in June 1940 to keep contact with the CWAECs, each assuming responsibility for four or five counties. Through them and the ministry officials, direct lines of communication were maintained at monthly conferences, which normally were also attended by the Minister, who briefed the meetings on the latest position nationally. The Liaison Officers became well known and formed a valuable two-way link between Whitehall and the farming community, helping to provide uniformity of operations among the CWAECs. The Liaison Officers were men of standing but this actually meant that they might combine this task with being the chairman of a CWAEC, giving rise to ever more charges of oligarchic control at the local level.

There was also a Public Relations Officer appointed, and an adviser with business experience to help with the placing of contracts for agricultural machinery, both at home and abroad. Otherwise any specialists were attached to local CWAECs rather than to Headquarters, and might be, for example, staff from the farm institutes, which were asked to suspend their teaching after the end of the 1940 summer term (Whetham, 1952: 46; Murray, 1955: 321).

It was also important for the decentralization that there was no intermediate body between Headquarters and the CWAECs. This has generally been regarded as a triumphant success, in part due to the personnel involved and the teamwork within the Ministry. In retrospect it was also seen as successful that the Ministries of Agriculture and Food had been kept separate, rather than being merged (as some had advocated). The Ministry of Agriculture continued to deal with the complexities of farming and the producers, whilst the Ministry of Food dealt with overseas food procurement, distribution and rationing among the consumers. A first Ministry of Food had been formed in the emergency of the Great War and had lasted from 1916 to 1921, when its remaining powers were transferred to the Board of Trade. In 1939 a second Ministry of Food had been created on the outbreak of war, with W.S. Morrison as Minister, and this lasted through the immediate post-war rationing years of 'victorious poverty' until the merger with the Ministry of Agriculture, announced in October 1954.

Interwar Agricultural Surveys

The interwar period was characterized by an enormous interest in the survey of agriculture, land use and farms. A large number of reports and maps were produced, and indeed a new profession, that of the agricultural economist, was born. The great enthusiasm for agricultural survey and improvement is reminiscent of that demonstrated in the late 18th century by the production of *General Views* and the *Annals of Agriculture* (Barrell, 1972: 64–97). The great difference between the two periods, of course, was that the interwar period was characterized by agricultural depression. This engendered the increasing interest by the State in the condition of agriculture, discussed above, which in turn encouraged a professional interest in the improvement of agricultural production (Stapledon, 1935). Although there was already a massive accumulation of information about agriculture in the form of the annual 4 June agricultural census, little information was available on a farm-by-farm basis that was of any practical assistance to agricultural advisers at a local level. The interest in farm surveys was paralleled by the burgeoning development of the regional survey movement and visions of a planned countryside (Matless, 1992), the implementation of the national Land Utilisation Survey – which was heavily influenced by the regional survey movement (Stamp, 1950: 3–4; Rycroft and Cosgrove, 1995) – and the organization by the Forestry Commission of forest and woodland censuses and surveys (Watkins, 1984, 1985).

The fashion for agricultural survey stems to a large extent from the pre-First World War funding of agricultural education and research. Agricultural economics lectures had been given at Cambridge from 1896 and the subject was also taught at some agricultural colleges. A key development was the establishment of the Agricultural Economics Research Institute (AERI) at Oxford in 1913. This was partially sponsored by the Development Commission and directed by Charles Stewart Orwin (Rogers, 1999). An early product of this institute was John Orr's survey of agriculture in Oxfordshire, which was published in 1916 (Orr, 1916). Research ground to a halt later in the war, but in 1919 the AERI was re-established with support from the Ministry of Agriculture. In this same period agricultural economics was taught as a separate subject at some universities, and in 1927 W.A. Ashby was appointed to the first chair in the subject at University College, Aberystwyth.

The engagement between farmers and the State in the First World War led to increasing interest in agricultural costing and farm accounting (Orwin, 1938: 23):

> The war gave a great fillip to agricultural costing. The control of the national food supply by the Ministry of Food raised the question of the proper prices for farmers' produce in an acute form. For some commodities, such as wheat, it was decided to restrict the prices paid to the home producers to figures below the world prices. For other commodities, such as milk, for which there was no world price, figures had to be guaranteed high enough to ensure the supply.

Prices were generally agreed at meetings between farmers' representatives and Ministry officials but just before the end of the war the Ministry of Food, with the Board of Agriculture, set up the Agricultural Costing Committee whose job was to collect agricultural costs to help to set prices. The country was divided into regions, for which a regional costings officer with accounting staff was appointed. This regional structure was disbanded in 1919, however, before it became fully established.

Work on costing continued at the Oxford AERI and at the Yorkshire Agricultural College, and in 1922 Orwin proposed to the Ministry of Agriculture that this work should be extended (Orwin, 1938: 17). In 1923 the first Advisory Economists were appointed at Wye, Reading and Cambridge. One of the recommendations in the final report of the Departmental

Committee on the Distribution and Prices of Agricultural Produce (1924) was that 'the Ministry of Agriculture should consider the advisability of obtaining funds to allow the appointment of a marketing advisory officer in each of the educational districts into which the country is at present divided' (Linlithgow Committee, 1924, in Orwin, 1938: 17). Following this recommendation advisory economists were appointed at university departments at Aberystwyth, Bristol, Newcastle and Manchester and at three agricultural colleges: the Midland Agricultural College (Sutton Bonnington), Harper Adams (Shropshire), and Seale Hayne (Devon). These new centres, together with those established earlier, provided agricultural advice and education for the 11 agricultural provinces (Astor and Rowntree, 1938). These provinces became of increasing importance in agricultural survey work and were later a key part of the organization of the National Farm Survey (NFS).

The growth in the number of professional agricultural economists stimulated the establishment of the Agricultural Economics Society in 1926. A wide range of research was carried out by agricultural economists through the late 1920s and 1930s. Much was concentrated on agricultural costing and accounts, but agricultural surveys, including work on the sugar beet industry, grass drying and milk production and also marketing surveys and social surveys, including FitzRandolph and Hay's (1926, 1927) extensive survey of the rural industries of England and Wales, were also carried out. These surveys were often funded by bodies such as the Development Commission and the Ministry of Agriculture. Many of the surveys were on a relatively small scale and there were increasing demands for national surveys. Orwin (1930: 120) argued that 'the essential preliminary to reconstruction is a national survey of agriculture'. He thought that this should entail a survey of farming systems and that such a survey would allow increased specialization in farm enterprises, the extension of farm mechanization and the rationalization of farm holdings (Orwin, 1930: 123).

One important development, influenced by the work of American agricultural economists, was the instigation of farm management surveys. The Milk Marketing Board funded a survey in 1934/35 by the Advisory Economists and the Oxford AERI 'to obtain general economic information about the milk industry and to use it for the benefit of the industry in advisory and other work' (Orwin, 1938: 38). About 60 farmers in each 'province' were sent forms by the Advisory Economists. Orwin (1938: 39) noted that 'an essential feature of the scheme was that copies [of the schedules from each farm] were to be sent to the Agricultural Economics Research Institute'. Here we see, in embryo, the sort of scheme that was later employed in the NFS with Advisory Economists being responsible for the collection of farm-level data for their provinces, subject to overseeing by a national organization.

This kind of farm survey was expanded in 1936 when the Ministry of Agriculture asked a committee of representatives from the AERI and advisory economists to establish an organization 'for the study of farming in all its principal types in England and Wales' (Orwin, 1938: 42; Whetham, 1978: 297). The survey was based on 1300 farms in 1936 and 1800 in 1937. Each of the provincial Advisory Economists was responsible for selecting about 200 farmers 'willing to furnish complete financial and statistical data about their farming' (Orwin, 1938: 42). The Farm Management Survey became an established annual study and, as Murdoch and Ward (1997) have noted, it not only provided national statistical information on the nature of the farm business, but also acted as a means of popularizing modern forms of farm accounting. At the instigation of the Farm Management Survey, Orwin (1938: 40) expressed concern that it was difficult to compare small farms 'operated in the main by family labour' with 'larger holdings on which most of the work is paid for in cash' and considered that 'It may be doubted whether it is possible to bring costs of production under a wage-labour system into line with those of a peasant industry' (Orwin, 1938: 41). Through its selection of a biased sample of farmers who could provide farm business statistics, 'the survey data that was … analysed yielded particular representations' of farming which became accepted as true (Murdoch and Ward, 1997: 315). The published survey reports presented to farmers, landowners and the agricultural policy makers a series of exemplar model farms that represented a reconstructed, modern agriculture and acted as a means of exhorting agricultural improvement. It is pertinent that Orwin (1938: 62) concluded a review of 25 years of agricultural economics research by setting it in the context of agricultural science:

> The Agricultural Economist's laboratories are the farm, the estate and the market. He has no attractive experiments, no crops to display under different conditions of growth, no animals, demonstration plots, or field plots. His work depends entirely on the collaboration of Government Departments … of the Marketing Boards and other official bodies; and of societies of landowners, land administrators, farmers and farm workers.

Agricultural economics developed quickly in the interwar period in conjunction with increased state interest in the control of agricultural prices. The importance of collecting detailed information about farm accounts in order to control production and set minimum prices was recognized in the First World War. After the war, research and surveys were carried out hand in hand with the development of a professional interest in farm modernization through specialization, mechanization and rationalization. However, it was not until a network of regional agricultural economists was established, each with their own province, that detailed information that could be used at the national level began to be collected on a regular basis. The Farm Management Survey made use of a network of individual farmers who provided detailed cost information and this had become well established by the end of the 1930s. The development of the profession of agricultural economics is closely tied to the increasing level of state intervention in agriculture during the interwar period, and was indeed largely funded by the

Ministry of Agriculture. It was not until the outbreak of war, however, that plans for a national survey of farms that had been nurtured during the inter-war period could be expedited.

By 1939 technical and scientific agricultural advice and survey had become firmly established in a tripartite structure (Whetham, 1952: 11). At the top the Agricultural Research Council coordinated the research at the specialized institutes and agricultural colleges, while at the 11 provincial centres there were advisers on local problems and monitors of local farming conditions. At another level the county councils in England and Wales also had advisory and teaching staff at their agricultural colleges. This relatively elaborate structure meant that at the outbreak of war there was a pool of local people who were well trained and with a knowledge of local farming.

The Development of County Agricultural Committees

The first County War Agricultural Committees were established in 1915, following a recommendation of the Milner Committee on agricultural output, in order to help to increase home food production through education and advisory work (Milner Committee, 1915; Murray, 1955: 9; Cox *et al.*, 1991: 40). They were established by the county councils, and often based on pre-existing county agricultural committees which had had responsibility for education or the management of agricultural land owned by the authorities. The new committees were frequently large, with representatives from many organizations: the Devon Committee had 38 members (Cox *et al.*, 1991: 40). These committees were supported by district committees which, again, could have many members. Murray (1955: 8) called these committees 'excessively large' and they were modified and given greater powers in 1917. Each of these 61 new CWAECs consisted of seven members from the county committees together with one or two members appointed by the Board of Agriculture (Murray, 1955: 8). Each committee had a chief executive officer, who was a full-time civil servant. This person was often the existing county organizer of agriculture (Whetham, 1978: 92). The large county committees continued to exist but 'soon sank into obscurity' as they did not have executive powers. District committees made up of 'respected farmers and local residents' were appointed by the CWAECs to 'carry the production policy on to every farm in every parish' (Whetham, 1978: 91). The membership of these committees was usually the same as for the district committees established in 1915 (Dewey, 1989: 181). Those people selected tended to be the larger farmers who employed labour, both because they were seen as the most successful and because they could spare the time required. They 'exercised a crucial role in advising the CWAECs and carrying out their instructions' (Cox *et al.*, 1991: 41).

The CWAECs' strengthened powers included the ability to make Cultivation of Land Orders under the Defence of the Realm Act 1914. Brown

(1987: 71) notes that they acted 'primarily through consultation and exhortation, but were given powers to order the recalcitrant to plough up pasture and to bring land up to a satisfactory standard of cultivation'. The exceptionally bad could even be dispossessed by the authority of the committee'. The old county committees, according to Cox *et al.* (1991: 41), had 'soon discovered the difficulty of performing a task of political control without [having] formal regulatory powers'. Various new specialist sub-committees dealing with issues such as labour, machinery and supplies were established and surveys of the supply of labour, fertilizers and seed in relation to the 1917 harvest were carried out (Murray, 1955: 8; Whetham, 1978: 99). The CWAECs were grouped into 21 districts, each having a Commissioner appointed by the Board of Agriculture. Thus the former loose organization of county committees was considerably strengthened, and the extra powers derived from the Defence of the Realm Act, together with the imposition of bureaucratic control by the Board of Agriculture, greatly enhanced the power of the state in determining local agricultural practices.

The CWAECs received their instructions in a series of circulars from the Board of Agriculture. An important development (Whetham, 1978: 92) was the requirement to survey all farms and to:

> classify them into those that were already well-farmed and could easily undertake an extra output of crops, those on which the farming could be intensified under supervision, and those which were badly farmed and which might need radical measures before they could contribute extra production of crops.

This survey work was carried out by the district committees. In its use of a threefold classification can be seen the origins of the farm classification system introduced during the Second World War and used in the NFS. The threefold classification was recommended by the Food Production Department of the Board of Agriculture and was modelled on a survey that had been carried out by the Essex War Agricultural Committee. This had divided farms into three types: well farmed, indifferently farmed but capable of improvement, and derelict. Each CWAEC was informed of the Essex method of survey and provided with suitable Ordnance Survey plans (Dewey, 1989: 172).

Much of the day-to-day work of the CWAECs and their sub-committees was concerned with issues of labour supply, exemptions from military service, ploughing, machinery supply, land drainage, horse sales, the supply of seed potatoes and the distribution of fertilizers (Dewey, 1989: 181–194). Although the CWAECs had the power to take possession of land, this was used only rarely. Whetham (1978: 99) points out that 'the success of the food production campaign depended on the goodwill of farmers and landowners; and the committees were so heavily burdened by their task that they ... were unwilling to farm land directly'. The powers to take over land were made particularly unpopular by the lack of any meaningful right of appeal (Dewey, 1989: 172).

The increased centralized power was largely vested in the new Food Production Department of the Board of Agriculture. This department was established at the beginning of 1917 to develop and enforce agricultural production policies, and it circulated to each CWAEC target acreages of corn for each county and areas of grassland that should be ploughed (Murray, 1955: 9). The President of the Board of Agriculture, Lord Ernle, stated that although 'ploughing was done under orders from the Executive Committees' the great majority of such orders 'were carried out willingly'. Moreover: 'In the vast majority of cases improvements [to cultivated land] were carried out by the occupiers themselves according to the directions of the Committees' but land held by 317 occupiers was 'placed under other management' (Lord Ernle quoted in Murray, 1955: 9). The Food Production Department was closed down in 1919 but some of its remaining functions were transferred to the new MAF in 1920. The CWAECs gradually had less work to do and were instructed 'to refrain from further control over the use of agricultural land' in the spring of 1919 (Whetham, 1978: 122).

Murray (1955: 16) considers two of the important legacies of State involvement in agriculture during the First World War were 'a strengthened Ministry of Agriculture' and 'the successful experiment of executing Government policy through voluntary committees composed largely of farmers themselves'. The *Agricultural Policy Sub-committee Reconstruction Committee* 1916 (Selborne Committee) had argued for the continuation of the system of county agricultural committees after the war. Such a system was maintained by the Ministry of Agriculture and Fisheries Act 1919 and initially the county committees had considerable importance in maintaining standards of husbandry and estate management (Dewey, 1989: 194; Cox *et al.*, 1991: 44). With the repeal of the 1920 Agriculture Act, however, their powers were much reduced, and they concentrated again on agricultural education, the management of council-owned farms and small holdings and 'the "policing" of certain forms of agricultural legislation' (Murray, 1955: 18; Wood, 1982). Under the Agricultural Holdings Act 1923, for example, a tenant had to be given compensation when given a notice to quit, unless the county agricultural committee issued a certificate of bad husbandry (Whetham, 1978: 218).

Although given a reduced role, some commentators saw the county committees as harbingers of greater state control. The Rural Report of the Liberal Land Committee 1924/25 suggested, for example, that the state should purchase agricultural land and that this should be managed by the county committees (Whetham, 1978: 218). Stapledon (1935: 256) felt that county committees should administer agricultural contracting and supply services. Moreover, the committee members continued to gain useful county-wide experience in agricultural issues. Cox *et al.* (1991: 46) point out that the 'NFU activists' who stayed as committee members gained experience which 'stood them in good stead when stringent powers were again assumed by County War Agricultural Executive Committees at the outbreak of the Second World War'.

A committee set up in 1935 by Walter Elliot reported in the following year and stressed the need for the establishment of county committees to encourage the conversion of pasture land to arable and to secure adequate supplies. These committees should be both executive and advisory. This report was accepted by the Food Supply Sub-committee of the Committee of Imperial Defence in 1936 (Murray, 1955: 49). By the end of the year a chairman, executive officer and secretary for each proposed committee had been selected and the Minister of Agriculture placed principal committee members on standby in the autumn of 1938 (Murray, 1955: 59). The Minister selected the executive officers for the committees from the professional agricultural establishment for each county. Most were established county land agents or county agricultural organizers.

Although linked to the county establishment, it would be a mistake to assume that the new CWAECs necessarily drew members directly from the existing county council agricultural committees. Murray (1955: 325) emphasizes that a crucial difference between the organization of the wartime committees in the two world wars was that the new committee members were

> appointed as individuals to act as the Minister's agents ... no organisations, not even the County Councils, were consulted about their appointments except in the case of agricultural workers on the Committee where consultation with the Trade Unions was inevitable. The result was that the Executive Committees owed loyalty to the Ministry and to the Ministry alone.

Although Murray emphasises the independence of the Minister's selection of members, however, it is clear that there was a fairly limited pool of people with the time, experience and ability suitable for membership of the new committees. Moreover, the unelected nature of the new committees was later the cause of considerable adverse criticism.

The Minister of Agriculture and Fisheries was vested with 'certain powers for the purpose of materially increasing home food production' under Regulation 49 of the Defence (General) Regulations (1939) and CWAECs were appointed under Regulation 66. The powers of the committees were very extensive, allowing them to take 'all necessary measures to secure that the land in their area was cultivated to the best advantage'.[4] The executive committees normally had from eight to 12 members, who were all appointed by the Minister, and it has been calculated that the total membership of the CWAECs amounted to 582 (Murray, 1955: 324). The committee members were generally chosen 'to organise the county and not to represent special interests' (Murray, 1955: 59). However, one member did represent the farmworkers and another the Women's Land Army. When the war began the committees were immediately set in motion, and they proceeded to administer the wartime farm effort throughout the war. The administrative staff of each CWAEC included a number of special officers, including district officer, machinery officers, drainage officers and so forth. These specialist officers were in addition to the clerical and administrative staff. Murray (1955: 326–327), the official historian of wartime agriculture, wrote:

> The speed and efficiency with which the whole organisation started to work were a source of amazement even to those who had been optimistic about its potentialities. Offices, clerical staff and equipment were hastily acquired, District Committees organised, County and parish statistics and maps distributed, and executive and technical staffs recruited ... . The rapid gearing of this elaborate machine to traverse the gap between Whitehall and about half a million different agricultural holdings was a magnificent achievement.

The clerical staff had to establish offices and assemble equipment, materials and maps very quickly. In addition to sets of Ordnance Survey maps the CWAECs were able to draw on the work of the Land Utilisation Survey. Stamp (1950: 15) noted that:

> An unexpected result of the declaration of war was the receipt of several telegrams from counties asking to borrow sets of original [Land Utilisation Survey] field maps as a basis for planning the ploughing campaign. ... the next two or three months were occupied in sending out thousands of our original six-inch field sheets on loan to County War Agricultural Committees ... our maps, of course, showed the areas actually ploughed in the years before the war.

The work of the CWAECs was carried out through a variety of local sub-committees, with the CWAECs themselves providing an overall integrating role. This gave a highly dispersed and localized structure to this example of state intervention – a factor that was to provoke both praise and concern as the war continued. Many of the general methods of workings of the new county committees were based on the experience of the equivalent First World War committees. Firstly, there were the district committees, what Murray (1995: 60) calls the 'most sensitive fibres in the "nervous system" connecting Whitehall and the individual farm'. These were made up of four to seven 'residents with a good knowledge of local farming, able and willing to give a certain amount of voluntary work, and also carrying the confidence of their fellow-farmers' (Murray, 1955: 59–60). The districts for which these committees were reponsible were usually organized on a Petty Sessional or Rural District basis and undertook, for example, the brunt of the National Farm Survey, using technical officers, district officers, personnel from agricultural colleges and voluntary help from land agents and farmers who knew local conditions of soil, climate and productivity very well. Secondly, there was a very extensive range of specialist sub-committees, which undertook functions connected with the training, supply and welfare of labour, the supply and transport of machinery, estate management of lands controlled by the committees, technical developments in husbandry, milk production and poultry, land drainage, farmers' supplies, cultivation and ploughing orders, pest and disease control, provision of cottages and, where appropriate, horticulture.

The CWAECs had enormous powers under the Defence Regulations (Murray, 1955: 326). They could take possession of land, requisition property, enter upon and inspect land, control the use of agricultural land and direct the cultivation of agricultural land. Much of the day-to-day work of the CWAECs was related to the encouragement of good agricultural practice. Monitoring the performance of individual farmers was also a key task and clearly the observation and recording carrried out as part of the NFS was an integral part of this scrutiny. Other tasks included: the allocation of county ploughing quotas; the payment of the £2 per acre ploughing subsidy; the distribution of tractors and other equipment; liaison with the armed services regarding labour supply; encouragement of land drainage; and the provision of mobile gangs of labour.

The range of work carried out by the CWAECs (and the post-war committees) is recorded in their minutes and those of their sub-committees and district committees. The PRO MAF 80 class of records contains minutes dated between 1939 and 1957, arranged in alphabetical order by English and Welsh county. In addition to the MAF 80 records a wide range of CWAEC documents and records is stored in other MAF classes. These include: CWAEC constitutions and finance (MAF 39); correspondence on the dispossession of tenants and the military use of land (MAF 48); the supply of farm machinery (MAF 58); and other general files, which include some individual farm files (MAF 145–149; MAF 157–182). Although a very large body of CWAEC documentation therefore survives in the Public Record Office (PRO), much of the more ephemeral material was destroyed.

The range and quantity of work done by the CWAECs was vast, and increased as the war progressed. In June 1940 the committees were strengthened by employing more technical staff. In addition, the agricultural colleges and institutes were closed by the Minister, freeing their staff to be used for advisory work and liaison between Whitehall and the CWAECs was improved by appointing liaison officers (see p. 20, above). Through 1941 and 1942 the administrative burden of the CWAECs continued to grow. Particular tasks included: the administration of feedstuffs and fertilizer rationing; the provision of accommodation for the Women's Land Army; the employment and supervision of farmworkers used for land reclamation and drainage; and the organization of pools of tractors and other machinery (Murray, 1955: 330–338).

Murray (1955: 339) describes the CWAECs as the 'greatest triumph' of the organization responsible for wartime agriculture. He considers that the key figures in the local organization were the 'progressive leading tenant farmers and farming landowners' on the committees and envisages their role in heroic terms:

It is impossible adequately to describe the devotion behind the long hours spent in visiting farms, field by field, by day and by night, in all seasons of the year; the infinite patience required in cajoling reluctant farmers to change their systems, and, often, in surmounting the suspicions and criticisms with which some farmers greeted the advice of their neighbours; the determination required to overcome the tedium of committee work and the weariness of form-filling and report-writing added to the continuous labour of running their own businesses.

Murray (1955: 339) also notes that with agriculture there was 'an almost crusading enthusisam to bring about a renaissance in British farming' which applied not only to farmers and landowners but also to agricultural 'educationalists, research workers, and administrators'. It can be argued that the war gave an opportunity for those academics, researchers and commentators who had for a generation been proselytizing the cause of a modern agriculture to transform farming on the ground.

Many county committees did the job very thoroughly, with a missionary zeal and well above the Ministry's stated standards, and the Ministry of Information portrayed the CWAECs as 'perhaps the most successful example of decentralisation and the most democratic use of "control" this war has produced' (Ministry of Information, 1944). But in face-to-face survey work human relations might become strained, since the committees acted as the 'eyes and ears' of Whitehall. This was particularly likely over the vexed question of the grading system applied to farmers' abilities by the CWAECs. The well-known progressive farmer George Henderson (1950: 50) was critical of the membership of CWAECs. He considered that some had 'failed in farming' and 'lacking the character and ability to become farmers, hate to see other achieve it', he wrote (Henderson, 1950: 98–99):

My own farm has been cited … as a model of production and efficiency … yet I was always depressed and angry whenever I was visited by the local member of the Committee in the early days of the war. What must have been the state of mind of farmers whose farms were not equipped and organised for an all-out production drive based on a balanced system of farming?

Opinion was divided as to whether the CWAECs had a future after the end of the war. Some commentators saw a modified form of county agricultural committee as crucial to post-war agricultural policy. Authors as varied as A.J. Hosier (the inventor of the milking bail), A.G. Street (the farmer and novelist) and Sir A. Daniel Hall (the agricultural administrator) all agreed on this point and linked an evolving county committee structure to some form of land nationalization (Hall, 1941: 228–230; Vesey-FitzGerald, 1941: 128, 134). Others, such as the progressive farmer Frank Sykes, stressed that the CWAECs were a creature of the war and would not survive peace (Sykes, 1944: 119–120):

The War Agricultural Committees are essentially Fascist in organisation. The State ... nominates a man in each county to be chairman of the committee; he, in his turn, nominates his county committee, and they, the local representatives. All this is very similar to the guild of the Fascist state. ... Farmers are aware of the weakness of the War Agricultural Committees. In such time as this, in the main, they are willing to submit to decisions, the wisdom of which, ordinarily, they would question. ... The moment the emergency is past, friction will make this system unworkable.

The CWAECs remained in existence until they were transformed into the County Agricultural Executive Committees under Part V of the Agriculture Act 1947. There is no doubt that they had proved a very important means of transforming agriculture during the war and certainly the NFS could not have been carried out without them. Yet Frank Sykes' views were borne out and the post-war committees were not popular. Several county committees were roundly criticized by J. Wentworth Day (1950: 40) as flourishing 'on bullying, petty feuds, jobbery, and favouritism'. He went on: 'Alone the diploma-ed, bespectacled Committee minions, so many of whom have never farmed for a living, or failed miserably when they tried, stand omnipotent, squandering money and petrol, and rendering no public account of either.' According to Street (1954: 43): 'after the war the British farmers, by their meek acceptance of the Agriculture Act [1947], betrayed Britain's country life for material security' and by sitting on the 'fascist' County Agricultural Executive Committees the 'yeomen of Britain' had become 'the yes-men of Britain' (Street, 1954: 37).

The Plough-up Campaign

Although a debate had begun before hostilities broke out about the respective merits of increasing home food production compared with massive stockpiling, the latter argument was supported by relatively few, especially since stockpiling would anyway have depended upon large imports, which could not be guaranteed in wartime. In fact there were only 16 weeks of supplies stockpiled at the outbreak of war (Martin, 1992: 39–41, 1999).

There was less debate over the optimal way to produce the greatest output of food in calorific value, than there had been in the First World War. It was now accepted that land devoted to producing feedstuffs for animals would have to be diverted into production for direct human consumption. This meant the ploughing up of grassland. It was calculated that 1 acre of arable under wheat would produce 2 million calories, or under potatoes 4.1 million calories, whereas under pasture 1 acre would produce 120,000 calories from meat or 450,000 calories from dairying. The argument was backed up by the grassland survey of Stapledon and Davies which had demonstrated that about 60% of the pastures were of *Agrostis* spp. or other relatively low-yielding grasses. Although reliance on the physical presence of grassland varieties as a guide to policy ignored the crucial role of management in extracting the best results from whatever grassland was available, the plough-up appeared to be self-evident as the best option.

The calculations convinced the Cabinet, who now opted for maximum calorific output rather than optimum nutritional standards, but then there were still questions to be settled relating to the timing of the start of the plough-up: should it begin before hostilities commenced? There were also questions concerning the time during which the stored fertility would last before the grassland had to be fallowed or reseeded; the number and type of animals to be retained to consume by-products and help to maintain soil fertility; the priorities for which crops should be encouraged; and the relating of these national considerations to the complexities and potentials of the very different farming localities (Murray, 1955: 3–47).

Preparations continued, but in March 1939 Hitler ordered the invasion of Czechoslovakia and war with Germany was now seen by most as inevitable. By April 1939 a new Minister of Agriculture, Sir Reginald Dorman-Smith, pushed through proposals for a payment of £2 per acre for permanent grassland (defined as at least 7 years old) ploughed before 30 September 1939 and reseeded, sown to an approved crop or fallowed. The payment amounted to a sum at least twice that of the actual cost of the ploughing operation, and was a significant contribution towards the £5 per acre average for ploughing, cultivation and liming. Such an outlay could be recouped within a single year. The campaign thus began about 6 months before the outbreak of war. The Minister also recommended the stockpiling of tractors and associated implements. Both proposals formed part of the Agriculture Development Act 1939, of which the most important proposal was arguably the provision for payments for the plough-up of permanent grassland. The 'Shadow' CWAECs were also put on standby to come into action immediately war was declared. During late 1939 the CWAECs had to organize their affairs very quickly. Office staff, organization, equipment and procedures had to be arranged, whilst at the same time certifying farmers' claims for the ploughing-up subsidy and dealing with myriad enquiries.

The target for the first season (1940) was an extra 2 million acres of arable land, amounting to about 10% of the pasture area in June 1939. This was to be sown with wheat where feasible, or otherwise with potatoes, oats, barley, beans, peas, rye or mixed corn. There was no actual directive at this stage as to which cereals to plant; in fact, profit levels of oats and barley were higher than that for wheat and nearly 50% of the new acreage was planted with oats. German military success in Europe then created the need for a further extension of arable in 1941 and another 2 million acres was scheduled to go under the plough in the autumn/winter of 1940/41. In November 1940 the Minister announced that the system of fixed prices, originally planned to be temporary, would now last until the end of hostilities (and 1 year thereafter). Again, for 1941 wheat was to be grown wherever possible, but now oats, beans, kale and roots for animal feed were encouraged, together with an extra area for potatoes. This was based on the ideas of John Boyd Orr and E.P. Cathcart, who worked on the newly created Scientific Food Committee and advocated a 'basal diet' for British citizens that would include fats, milk and other vegetables as well as wholemeal bread – a diet that might be sustained through possibly years of siege (Hammond, 1954: 31–35). More than 4.25 million acres had therefore been ploughed up in the first two seasons, despite unfavourable weather, and there were now significant increases of oats, wheat and potatoes over the pre-war norms.

Stapledon had rushed to produce *The Plough-up Policy and Ley Farming* in June 1939, dedicated 'To all pioneers who had embarked upon ploughing up derelict grassland before the announcement of the government subsidy' but aiming to advise those who had not yet done so to increase 'emergency fertility' in this way. Moving from the Welsh Plant Breeding Station at Aberystwyth, he became Director of the Grassland Improvement Station in 1942 with farms at several differing locations – including Dodwell-Drayton, Warwickshire, which, after its establishment in 1940, proved a considerable influence on the adoption of ley farming in that locality (Woods, 1994: 18–19). Later Stapledon also published *The Way of the Land*, which included praise for previous governments in emphasizing agricultural research and education, for introducing the sugar beet subsidy, and for the more recent subsidies on lime and basic slag and for ploughing up old grassland (Stapledon, 1943: 254–255):

> It was to mean a great deal that the ploughing up of grassland had started before the war, even if only a short time before. Experience was gained and interest was created … and we can perhaps gain comfort from the fact that we were given a slightly better start this time.

Stapledon argued that there had been a psychological change since the Great War. Then, he argued, the acreage that could have been contemplated for ploughing up was limited to that which had lapsed from cultivation; now the plough-up extended into rough grazings and elevations even above the 1300 ft contour. The limits this time were those of machinery supply and the availability of phosphate fertilizer. He retired from the Grassland Research Institute in 1946.[5]

In fact plans for the 1942 season were laid to replace imports as far as possible (wheat, potatoes, sugar beet) or to produce those that could not be imported (fresh vegetables and milk) and a further enlargement of the arable area was required. However, it could already be seen that the stored fertility would soon be exhausted, and decisions had to be made about the timing of any maximum saving of imports. Eventually it was decided that another 1.25 million acres of tillage was required, and although the actual area of plough-up reached another 1.5 million acres, 300,000 acres reverted to temporary grass. Since there was also a continuing take-up of land for airfields and military uses, the net increase in the arable area was less than 1 million acres in 1942 (Hammond, 1954: 177; Foot, 1999). In late 1942, at a time when the import crisis was grave and food cargoes were being lost at a crippling rate, it was decided that an additional 1.1 million acres was needed for the 1943 season; although this was achieved, only a net 800,000 acres was added. In 1943/44 another 500,000 acres was ploughed but an almost equal area went back into temporary grassland.

The policy of maximizing the arable acreage persisted throughout the war, with a steady increase in the area of grain through to 1944, when the pace slackened off and farming interests began to review the damage done through years of continuous arable cropping. The acreage of oats had actually peaked in 1942, that of wheat in 1943, and potatoes in 1944. Cropping of cereals for 2–3 years in succession had now led to exhaustion on the lighter soils and weed infestation on the heavier clays. In 1945 the area of permanent grass, which had been cut back annually since 1939, rose again (Table 2.1). Thanks, too, to relatively favourable weather during the war years until 1946, yields of the main crops were above average, even though such increases have been questioned as being more a reflection of the low-yielding state of agriculture before the war (Martin, 1992: 101).

However, the continued consumption of animal products by the British people was due in large measure to the Lend–Lease agreement which ensured supplies from July 1941, to September 1945 from the US through imports of canned pork and fish, canned milk, dried eggs and fruit, as well as machinery. At home production of milk, beef, veal, sheep, pig meat and eggs all declined, the latter two very dramatically (Hammond, 1954: 70–72, 219).

This unprecedented change in British land use in such a short time – a change that Murray claimed as 'successful far beyond the calculations and estimates of pre-war planners' – was based on quotas given to each of the counties of England and Wales and enforced by the CWAECs. All but eight counties met their targets in the first years, and these counties were broadly those with large proportions of heavy soil, such as Essex and Buckinghamshire. Broadly, the brunt of the plough-up was borne by the pastoral areas. In 1939 the arable area of Leicestershire totalled 15% of the total area; by 1944 this had risen to more than 45%, and the situation in Nottinghamshire, Derbyshire and Staffordshire was similar. The lack of familiarity of such farmers with large-scale arable production made their success all the more striking, and on the small grassland farms the purchase of the necessary machinery would bring diseconomies of scale, since the CWAECs unremittingly pursued their quotas to big and small alike. In the farm economics of the late 1930s, the minimum viable size of a holding for cereal cultivation was 120 acres, and such farms often also contained many small fields which rendered mechanization even more problematic. Undoubtedly, such farmers fared worse and made losses compared with their large-farm counterparts in the arable eastern counties (Martin, 1992: 168–185).

Table 2.1. Crop acreages and livestock numbers on agricultural holdings in the UK 1936–1945 ('000 acres or '000 head).

Crop/livestock	Pre-war[a]	1939	1940	1941	1942	1943	1944	1945
Wheat	1,856	1,766	1,809	2,265	2,516	3,464	3,220	2,274
Barley	929	1,013	1,339	1,475	1,528	1,786	1,973	2,215
Oats	2,403	2,427	3,400	3,951	4,133	3,680	3,656	3,753
All grains	5,301	5,305	6,827	8,276	8,782	9,560	9,393	8,765
Potatoes	723	704	832	1,123	1,304	1,391	1,417	1,397
Sugar beet	335	344	329	351	425	417	431	417
Turnips and swedes	782	712	746	837	858	830	820	814
Mangolds	227	216	231	267	269	286	308	308
Flax[b]	23	23	65	128	118	145	184	124
Vegetables	278	291	304	375	422	423	504	512
Fallow	422	374	306	219	280	240	231	347
Total crops[c]	8,907	8,781	10,455	12,686	13,635	14,509	14,548	13,849
Temporary grass	4,181	4,125	3,891	3,553	3,863	4,219	4,725	5,334
Total arable	13,088	12,906	14,346	16,239	17,498	18,728	19,273	19,183
Permanent grass	18,750	18,773	17,084	15,114	13,706	12,330	11,735	11,840
Total agricultural area	31,838	31,679	31,430	31,353	31,204	31,058	31,008	31,023
Rough grazings	16,476	16,539	16,639	17,003	16,959	17,119	17,034	17,229
Dairy cattle	3,943	3,885	3,957	3,988	4,199	4,323	4,373	4,343
Other cattle	4,732	4,987	5,136	4,952	4,876	4,936	5,128	5,273
Sheep	25,785	26,887	26,319	22,257	21,506	20,383	20,107	20,150
Pigs	4,466	4,394	4,106	2,558	2,143	1,892	1,867	2,152
Poultry	76,236	74,357	71,243	62,059	57,813	50,729	55,127	62,136

[a] Average for 1936/37 to 1938/39.
[b] Flax for fibre.
[c] Total crops other than grass.
The table omits other fodder crops (e.g. cabbage, rape, vetches, etc.) and the residual 'other crops' of the June Agricultural Census.
Source: Murray (1955, 373, Appendix Table IV).

It has been shown that the CWAEC plough-up orders were formalities in many cases, and often issued only after the CWAEC officials had discovered that a field had already been ploughed in order to make the farmer eligible for grants or to protect him from his landlord, but the zeal with which most CWAECs pursued their targets has been a frequent source of comment. The plough-up campaign was a political, emergency policy – it was not an economic programme. Many farmers responded immediately to the £2 subsidy; many others felt that they were being ordered to plough up land that was uneconomic or that offended their sense of good husbandry.

Local relations were mostly harmonious, yet herein lay the seeds of local discontent, nepotism, the settling of scores, and consequent evictions from part or whole of farmland by the CWAECs. More than 440,000 acres were taken over and the tenancies of another 225,000 acres terminated, although cases where both land and farmhouse were taken over amounted to fewer than 1400 (compared with a total of over 300,000 holdings). Most cases of dispossession dealt with non-resident occupiers, fields rather than farms, or formerly non-agricultural land, but some evictions were poorly handled and there was no route of appeal. The media were generally reluctant to criticize the war effort, and the NFU, which had been instrumental in setting up the CWAECs, was thereby not neutral, and indeed could be seen as being in liaison with the government as part of a 'policy community' typified perhaps by Dorman-Smith's appointment as Minister of Agriculture at the outset of the war – a man who had recently finished a term of office as President of the NFU (Smith, 1990).[6] In certain instances farmers were issued with several compulsory orders, which put them under psychological pressure to leave. In the overall context of the war, mass opposition was impossible, but membership of farmers' groups in various ways opposed to the evictions grew. At its peak the Farmers' Rights Association numbered between 5000 and 6000 members, publishing books such as *Living Casualties* (c.1945), *The New Morality* (1945) and *The New Anarchy* (1948); the Farmers and Smallholders Association opposed all state subsidies and controls; while local protest groups such as the Essex Farmers and Countrymen's Association also grew (Whetham, 1952: 47–48; Murray, 1955: 302–303; Self and Storing, 1962: 111–113).[7]

War Policies

From 1 September 1939 British agriculture was formally put on a war footing. State control over agricultural production remained at a very high level throughout the war, mediated through the CWAECs. By mid-1943 control had reached its zenith. The new governmental structures propelled forward by the war did not always prove easy of bureaucratic handling. The Ministry of Food, formed in September 1939, and the Ministry of Agriculture – one responsible for the consumer's interests, the other for the producer's – had to coexist, and the former had evolved in turn from the Food (Defence Plans) Department. In 1939 it had been anticipated that the war would last for 3 years. Its prolongation served both to implant more firmly the machineries of government and to ensure the need for a continuation of emergency measures in the growing awareness that Britain would be impoverished by the close of the hostilities. It was also assumed that there would be a world food shortage following the war, and that the government would remain committed to welfare schemes such as the National Milk Scheme. The Ministry of Food therefore saw a continuing role in the transition from war to peace.

The translation from national policy to ground-level implementation was through the CWAECs, which administered government aims for farmers by ensuring that, as in the First World War, compulsory cropping orders were imposed, derelict land reclaimed, and some farmers evicted (Dewey, 1989: 171–197). Large numbers of cattle, pigs and poultry were slaughtered, all of which actions demanded full records being kept. Thus, within the wider remit of state intervention, or rather what J.A. Venn and other contemporaries chose to call 'control', the requirements of bureaucracy began to loom large (Venn, 1939: 21–49). Indeed, there was a renewed appreciation of the role of the local provision of agricultural education and record keeping. In the immediate pre-Second World War period there was an increasing degree of organization amongst the county council Agricultural Education Committees and Departments of Agricultural Education regarding their systems for maintaining a record of the farms in their county areas. Hampshire Agricultural Education Committee, for instance, had a highly developed system that made use of detailed cards for each farm.[8]

On 2 June 1939 the MAF issued a circular letter to local authorities for agricultural education drawing attention to the Hampshire scheme and stating that, in the event of war, such a scheme would be of 'great assistance'. It was hoped that similar schemes could be started by all counties. Other counties then immediately became keen to report on the record systems they maintained. On 27 July the Ministry sent a circular to the County Agricultural Organizers asking them to liaise with the Provincial Agricultural Advisory Centres with regard to the information that should be included in the county farm record systems. A similar circular was sent to the Provincial Advisory Centres. Several county committees resolved to implement a farm record scheme, and the Hampshire committee (reconstituted as a CWAEC after the declaration of war), thinking that the Ministry required a detailed Land Fertility Survey by the end of December 1939, put forward a proposal for their county using a staff of 40. The Ministry turned down this scheme and said that what was required at present was for CWAECs to begin the process of farm assessment and record keeping in order to assist with the plough-up campaign and the allocation of agricultural resources. This should be fitted in with their normal activities. Further instructions would come from the Ministry in the near future.

The idea of a complete survey of farming within England and Wales met with great ideological opposition in the later 1930s, despite the precedent of the Farm Management and Land Use Surveys, and the growing statistical contexts described in Chapter 1. However, there had been considerable pressure from agricultural economists for a national survey of farming land for some time, and in 1938 the Ministry of Agriculture and Fisheries considered a memorandum prepared by R.R. Enfield, an Assistant Secretary at the Ministry, on a proposed survey that incorporated the ideas of Professors Abercrombie and Engledow.[9] The memorandum was pigeon-holed, with the Ministry's comment that such surveys could only be justified 'if the government were to contemplate a measure of government control on Socialist or Germanic lines over the utilisation of agricultural land and the operation of individual farmers'.[10] Clearly there were to be limits to the degree of control over production and over the working lives of Britain's farmers. It is interesting to note that even after the National Farm Survey was instituted there were still those, like C.S. Orwin, who continued to put forward plans for massive state surveys of agricultural land and of rural communities in general, as he did in *Speed the Plough* (Orwin, 1942), and indeed as he had previously done in *The Future of Farming* (Orwin, 1930). Both Orwin and the highly respected A.D. Hall were advocates in effect for a nationalization of farmland in Britain, and the requirements of a full survey were thus linked to this ideology.

Only with the establishment of the CWAECs, and with the very real wartime need to increase food production and to ration fertilizers and animal feed, were the conditions right to initiate such a survey. Central government organization and initiative was necessary for the implementation of a national farm survey, although a large amount of land-use information had already been gathered pre-war and was drawn upon by the CWAECs.

The war cemented the concept of guaranteed prices within the farming community, even though it carried with it the necessity of State control. Expansion and a new confidence in the importance of farming had replaced stagnation, and home-produced food had become a more entrenched requirement. When the Labour Government assumed power in 1945 it found most of its priorities in place and largely acceptable to the agricultural interest. Stability and efficiency then became the central planks of the Agriculture Act 1947 (Self and Storing, 1962: 20–24).

Agriculture and the State, 1919–1945

In retrospect it is now evident that the preparations necessary for national defence as war loomed closer in 1939 were dependent upon an increased measure of corporate control within the agricultural sector accepted by both main political parties, by the NFU and by the industry at large in the inter-war years. This acceptance, allied with a growing expertise in rural surveys coordinated at a national level, and an important degree of responsibility delegated to the local level, was sufficient to see Britain through the war with her food supplies just about sufficient to meet the needs. This top-down approach was all the more necessary since the ability shown by the agricultural sector to cooperate within its own ranks was less impressive, exhausted perhaps by the years of depression prior to the late 1930s. Even after the outbreak of war, cooperation could not be guaranteed, and although most responded to wartime needs with alacrity, there were others (whose numbers cannot yet be ascertained with precision) who found themselves victims of local vindictiveness or who abused their new-found powers to pay off old scores.

State intervention after 1945, with its guaranteed markets and prices, therefore rested on a perceived need to ensure food production and supply in the straitened post-war years, as well as on a recognition of its success in pulling a depressed sector from virtual bankruptcy to full productivity within less than 10 years. That recognition, and the place of the CWAECs within the overall policy, ensured the continuation of local committees within the Agriculture Act 1947.

Notes

1   PRO MAF 38/80. *Notes on a Memorandum on Agricultural Policy by Lord De La Warr* (1939).
2   For further material on government policy at this time see A. Webber (1982) 'Government Policy and British Agriculture 1971–1939', unpublished PhD, University of Kent.
3   *Agriculture in the Twentieth Century* in many ways epitomizes the emphasis on scientific agricultural research by 1939, including as it does essays by A.W. Ashby on agricultural policies, J.A. Scott Watson on husbandry, C.S. Orwin on farm businesses, Sir E. John Russell on soils, Stapledon on grassland, Sir John Orr on national health, and many others.
4   PRO Introduction to MAF 80 Agricultural Executive Committees: Minutes, April 1972.
5   Sir George Stapledon (1882–1960) CBE, FRS, was the first Director of the Welsh Plant Breeding Station from 1919 to 1942. See also Fig. 7.9.
6   Reginald Dorman-Smith (1899–1977) was President of the NFU (1936/37), Minister of Agriculture (1939/40) and Governor of Burma (1941–46).
7   Martin has noted that Murray was actually more critical of the actions of the CWAECs than was evident in the official history, but these criticisms were deleted by the Assistant General Editor to the series, W.E. Hancock (Martin, 1992: 267). For material relating to the writing of the official history by Murray, including a typescript, see PRO CAB 102/27–30. The more critical unpublished narrative by E.H. Whetham, whose work formed the basis for much of Murray's book, is 'Agricultural Policy and Food Production' in PRO CAB 102/325–7.
8   PRO MAF 38/469 includes an example of one of the cards.
9   Sir Ralph Roscoe Enfield (1885–1973) had been a career civil servant since 1913, and served with the Ministry of Agriculture from 1919 to 1952. He became an Assistant Secretary in 1936, Principal Assistant Secretary in 1942, and Chief Economic Advisor in 1945. See also his book *The Agricultural Crisis 1920–23* (1924).
10 PRO MAF 38/206 *Notes on Proposals for an Agricultural Survey.*

The Organization of the National Farm Survey 1941–1943

3

In his authoritative survey of wartime food and agriculture in Britain, Hammond (1954: 32) notes that:

> only by a laborious survey involving one or more visits to every holding in England and Wales, could complete data be secured on which to plan for increased agricultural output. This survey and a somewhat less exhaustive one for Scotland were begun in June 1940 and completed two years later. They provided the information with which the greatest efforts of wartime production were organised.

However, Hammond has conflated what were in fact two distinct surveys for England and Wales: one begun in 1940, and one (the National Farm Survey (NFS)) begun in 1941. Whilst this book concentrates on the latter, it is none the less important to recognize the importance of its wartime forerunner.

The First Farm Survey 1940

Circular 227 from the Ministry of Agriculture and Fisheries (MAF), issued on 6 June 1940, initiated the first Farm Survey. The circular set out the various categories of data that were required, and instructed the County War Agricultural Executive Committees (CWAECs) that the information gained should be summarized by districts and by counties.[1] The principal concern of this survey was to increase food production, and the investigatory work was supposed to be completed before the end of July. The survey contained questions that were not included in the later Primary Record of the NFS, such as the method of drainage, the fertilizers needed, labour required (by type), vacancies for the Women's Land Army and accommodation, and livestock reductions in pigs and poultry. In addition, Circular 242 dated 12 June 1940 asked that information be gathered on tractor availability. On 21 June the Editor's Diary of the *Farmers Weekly* contained the following:

> This new Domesday of Britain is about the best thing that has happened for a long time. In peace it would have been worthwhile; in war it is essential ... I hope the committees will allow their courage and, indeed, their patriotism to suppress their sentimentality, and take strong action whenever needed.

Similar coverage followed in subsequent editions of the *Farmers Weekly*, with visits to Devon and Lancashire to witness the survey in action being given very positive treatment. The surveyors were praised for their hard work, and it was claimed that R.S. Hudson, 'in ordering the survey ... has perhaps done more for agriculture and agricultural education in the last three months than was done in the last three decades'. There is some suspect material: on 26 July a farmer, Mr R. Whittaker of Cockerham, wearing clogs and with socks tied over his knees for weeding turnips, greets the surveyors, 'What is your quest this time?' By October, the only anxiety was that the results would be 'consigned to County council lumber rooms'.[2]

To organize the results of this first farm survey and to plan for an extension of the survey (which was shortly to result in the NFS), a Farm Survey Committee was established, which held its first meeting on 9 October 1940.[3] By that date the Ministry was already planning for the farm survey to be developed as a 'Second Domesday Book' and setting out the proposed scope of the expanded survey – to be known as the NFS.

By October 1940 the first survey had still not been completed. Moreover, it had been carried out to greatly differing standards amongst the various counties. Pembrokeshire was noted for the fullness of its recording, and some counties, including Cheshire, used very detailed forms.[4] On the other hand we have the evidence of George Henderson (1944: 159) from the Oxfordshire Cotswolds that:

> In the early days of the war, after the farm had been inspected, we were told we were making the best use of the land, and no orders were made. Then the following year [1940], at the time of the great survey, which actually took nine minutes and consisted of writing on to a map the crops I said were in the fields, I mentioned that we intended to plough up fifteen acres of ley.

Circular 312 from the Ministry dated 3 August 1940, asked CWAECs to submit progress reports on the survey, and requested eventual summaries of the information asked for in Circular 227. The information gained was primarily for the benefit of the CWAECs themselves, in particular in the identification of 'C' farms and as a source of stimulus for corrective action to improve food production.

Circular 416 dated 17 December 1940 from the Ministry thanked the CWAECs and their district committees for their hard work in carrying out the survey.[5] It also repeated the request for a summary of the results that they had obtained which were to be organized in a manner that was indicated on the reverse of the circular. When these returns came in it was found that they varied considerably. Some used (literally) the reverse of Circular 416, while others (e.g. Lancashire) were immensely long and included an analysis of the data. Ten counties, in fact, appear never to have submitted their returns, but a summary report on the survey was prepared without them.[6] It was this inconsistency of the 1940 material – the need to obtain conformity of data and to eradicate the obvious errors and anomalies thrown up by the 1940 survey – that provided one of the main reasons for the later NFS.

Circular 416, in fact, had already stated that a survey would now be developed in a manner more standardized than the first attempt, and that this greater survey would not only help with the immediate food production programme but would provide additional information to assist post-war agricultural planning. The 'uniform form of record' that was required would be put together 'from the information already obtained by Committees', and the staffs of the Advisory Economists would assist in its preparation.

The Organizational Structure for the Implementation of the National Farm Survey 1941–1943

The arrangements for the successful completion of the NFS were necessarily complex, and involved several different branches of the State apparatus connected with the land. The main organizations involved, and their roles, are therefore listed briefly here, in order that reference can be made to them later in this volume.

Ministry of Agriculture and Fisheries (MAF)

The Head Office was at 55 Whitehall, London SW1. The Ministry had further London offices, including 15 Whitehall and 23–25 Soho Square. During the war the offices of the Statistical Branch were at St Anne's-on-Sea, Lancashire (actually in the Hotel Majestic and Hotel Lindum). The Minister of Agriculture and Fisheries at the time of the NFS was Robert S. Hudson, who had been so appointed to serve within Churchill's coalition government in May 1940. His Permanent Secretary was Sir Donald Fergusson and the Assistant Secretary (Economics and Statistics Division) was R.R. Enfield[7]. Within the Ministry there were a number of Principals: W.R. Black (Research, Technical and Publications

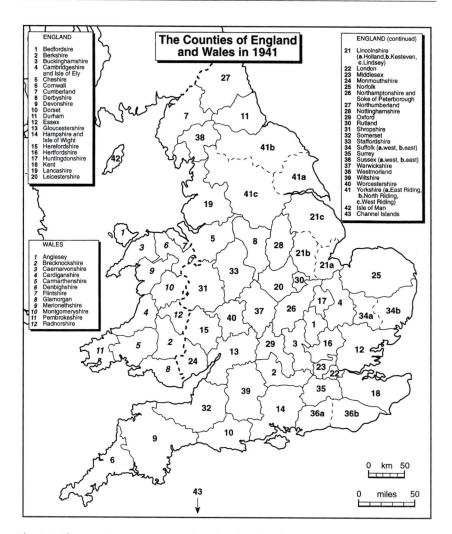

Fig. 3.1. The constituent counties of England and Wales. Each had its own County War Agricultural Executive Committee (CWAEC).

Branch), M.G. Kendall (who was later appointed to the Chamber of Shipping in July 1941), J.H. Kirk (Economics Branch) and two 30-year-olds, A.B. Bartlett (Statistical Branch) and H. Whitby (Economist)[8]. The division of the Ministry responsible for the NFS was the Statistics and Economics Division. Kendall, Kirk and Whitby were themselves commentators on the farming scene in several varied ways (Kendall, 1939; Whitby, 1946; Kirk, 1979).

The separate Ministries of Agriculture and Fisheries on the one hand and of Food on the other were amalgamated in April 1955.

County War Agricultural Executive Committees

The CWAECs were formed on the outbreak of war, but had been in preparation from 1936. There were 62 county committees in total, with 49 in England and 13 in Wales (Fig. 3.1).[9] The CWAECs were made up of eight members, including the Chairman, who had been selected (even before Munich) because of his agricultural expertise by the Lord Lieutenant and the Land Commissioner. These were to be the 'visible human chain' between Whitehall and every farm in the country (Ministry of Information, 1945b). The Chairman was then virtually given a free hand to select his committee, which was normally composed of leading farmers and landowners, many of whom now set about restoring the land, after years of neglect, with a crusading fervour. Attached to the committee (the members of which served unpaid) were salaried personnel. There was usually a full-time Executive Officer – normally either a county land agent or from the staff of one of the agricultural organizations of the relevant county council, with necessary technical knowledge but who might normally have experience in local government as well as agriculture. The committee was completed by the Secretary and the Land Commissioner. Meetings were held weekly. Chairmen had as much independence to act as they thought appropriate for their counties: Lord Cornwallis, chairman of the Kent WAEC, actually circulated an unofficial survey to all Kent farmers of land over 50 acres in May 1939, asking them for estimates of the amount of grassland that might be ploughed up in the event of war, how much was then being ploughed under the subsidy scheme just announced, and how much labour would be available. The Kent district organization and a shadow CWAEC were also decided by that date (Cox, 1944: 119; Burrell et al., 1947: 76).

The CWAECs also had sub-committees and district committees. The sub-committees served to deal with aspects of the Committee's ongoing work such as horticulture, labour, machinery or livestock feedstuffs. The areas represented by the district committees, of which there were 478 in total in England and Wales, corresponded generally with those of the rural district councils, and each district committee usually had from four to seven volunteer members who were local residents. At any one time there might be a total of 4652 people engaged with these district committees, supervising an average of 5000 acres per committee (Easterbrook, 1943: 25–26; Martin, 1992: 81–83) (Fig. 3.2).

The survey work necessary for the National Farm Survey was carried out by the district committees of the CWAECs. As Henry Williamson (1967: 19) wrote from his East Anglian farm:

> In the New Year of 1941 there was some talk of making another Domesday Book – a complete survey of the land and stock of Great Britain. One afternoon a tired woman from the War Agricultural Executive Committee arrived at the open farmhouse door.

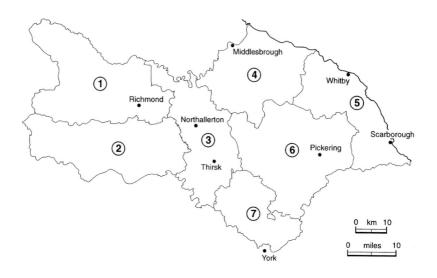

Fig. 3.2. County War Agricultural Executive Committee (CWAEC) district committee boundaries in the North Riding of Yorkshire (adapted from: PRO MAF 38/213).

The MAF, in its Circular 545 of 26 April 1941, had told CWAECs to submit details of extra staff required to carry out the survey. Some CWAECs were determined not to ask for extra staff while others simply could not find suitable additional personnel for the purpose. Somerset CWAEC asked for 24 surveyors (three to each of eight districts), and had eight approved at a cost of £250 p.a. each. Wiltshire put in a bid for one coordinating officer and six district officers at salaries of £250 p.a., all of whom were approved.

The surveyors needed a firm technical and professional knowledge of farming to carry out their survey work effectively. Some CWAECs used their existing District Officers, Technical Officers, or members of advisory staffs and personnel from agricultural colleges for the purpose. Others took on special staff, such as land agents, but they also used voluntary, unpaid help from farmers, Crop Reporters and even university students. Some CWAECs sought not to use farmers for the survey work, but most were forced to do so, owing to the lack of available and suitably skilled personnel. When farmers were engaged, care was supposed to be taken to ensure that they were not reporting on their own farms or those of their near neighbours, but as an official source put it (Ministry of Information, 1945b: 13):

> Hundreds of field-workers, mostly volunteer Committee-men or retired farmers, began the gigantic task of surveying every holding with more than five acres of land. They covered every shire and parish; they worked with 6-in

scale ordnance maps, tact, circumspection, and plain physical stamina. For they had not only to assess the qualities of the land, they had to sum up the qualities of the farmer himself.

Less charitably perhaps, the foreman Bert in Edward Blishen's *A Cack-handed War* (Blishen, 1972: 38–39):

> took a simple view of War Ag officials. They were all failed farmers, or opportunists with dubiously relevant backgrounds who had wormed their way into their indefensible jobs. … It was anger, really, at the invasion of his world by men clutching papers. 'More bloody papers,' he'd say, as one of the little vans hurried importantly up the chase.

Certainly, as the war drew to a close, there were many who felt that the district committees were assuming effective power and that the executive committees were merely rubber-stamping district committee decisions on such issues as ploughing orders, for example (Bateson, 1946: 159–160).

Advisory Economists and Advisory Centres

As government intervention in agriculture increased during the 1930s, the role of agricultural economists had developed proportionately. Systematic provision of information was now required to guide policy and agricultural economics as a science had now come of age. Departments of agricultural economics at universities were now funded directly by the Ministry, with a somewhat narrow technocratic and practical orientation as a result (Cox *et al.*, 1986: 1–19). There were 11 Agricultural Advisory Centres in England and Wales, each with a Principal who attended the Conference of Advisory Economists (Fig. 3.3). These periodic conferences, with a membership of some 15, debated questions arising from the NFS. A sub-committee of the Conference was also established (with nine members) in October 1940 to consider aspects of the NFS, including the questions to be asked on the Primary Return form. A further sub-committee met on 6 February 1941 to consider staffing arrangements for undertaking the Survey (see Appendix).

The Advisory Centres maintained professional research staffs working on technical aspects of agricultural science as well as the economics of farming, with financial support from the Development Fund.[10] The Advisory Economists also administered the Farm Management Survey, which had begun in 1936 and which each year produced data from 2000 holdings in England and Wales. The Agricultural Advisory Centres represented the 'think tanks' of British agriculture, and were able to keep the Ministry of Agriculture informed on regional problems and the latest ideas and research.

For their role in the NFS, the Advisory Centres were financed by the MAF in order to help them find the additional staff required. Each of the 11

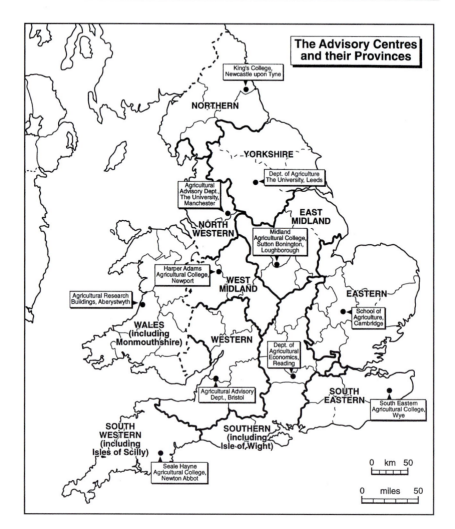

Fig. 3.3. The Agricultural Advisory Centres and their Provinces.

Centres was to have one Farm Survey Supervisor (at a salary of £340 p.a., although the original proposal was £400). The supervisor was to liaise with the CWAEC district committees, which would involve frequent travel to their offices. For this purpose, a travel allowance of between £150 and £250 p.a. (depending on the area) was allowed. Some supervisory positions were filled by moving the Advisory Centre's permanent Technical Assistant into it and replacing him with another man appointed in his place. Ideally, but not essentially, a supervisor held academic qualifications in agriculture. Assistants to the supervisor (who were to be known either as Field Workers or Recorders) were to be allocated at a rate of one assistant to every 10,000

Table 3.1. Numbers of holdings covered by Farm Advisory Centres and numbers of assistants.

Centre	Number of holdings	Number of assistants
Aberystwyth	50,000	5
Cambridge	38,000	4
Bristol	31,000	3
Leeds	30,000	3
Midland	28,000	3
Seale Hayne	25,000	2 (originally 3)
Reading	23,000	2
Harper Adams	22,000	2
Manchester	21,000	2
Newcastle (King's)	18,000	2
Wye	17,000	2
Total	303,000	30

holdings for which the Centre was responsible and at a salary of £200 p.a. each. The number of assistants required can be seen in Table 3.1.

It was said that another 10,000 holdings were 'likely to be found', but that the work on these would have to be absorbed without additional staff. There was clearly some variation in the ratios of assistants to holding numbers between Centres. Wye was obviously in a much better position (with one assistant for 8500 holdings) than Seale Hayne (one assistant for 12,500 holdings).

As with the supervisors, the assistants (Recorders) would also have to travel to the district offices of the CWAECs (where a great deal of their work would be done), and a travelling allowance of £50 each a year was allowed. It is not clear whether the Recorders needed to stay overnight in the towns where the district offices were situated. It is mentioned that they were to be stationed initially at the towns where they were to work, but also that they should be recruited in the town where the Advisory Centre was situated. In certain Advisory Centre provinces it seems that a great deal of work with the CWAECs was done by correspondence rather than by personal visits to the district committee offices.

Each Advisory Centre was also to have a filing clerk, who would arrange the records as they arrived at the Centre and were completed, and create a file for each holding, in which both the Primary Returns and Census Returns would be kept. The files were to be indexed through a card index system which would enable the file of a particular holding to be found quickly. From September 1942, it was decided to add additional documents relevant to a particular farm to these files (now termed dossiers), but making sure this was not confused with the NFS material. These additional documents came from other work of the Advisory Economists, such as the results of investigations into the use of fertilizers or the prevalence of wireworm. The filing clerks would also deal with correspondence with the CWAECs. The original proposal was that each filing clerk would receive an annual salary of £250 a year, but in May 1941 the Advisory Centres were advised by the Ministry of Agriculture to recruit women for these posts at the cheaper rate of £150–200 p.a. (plus bonus). This economy cut the total amount of anticipated expenditure (excluding equipment) for the Advisory Centres in the first year of the Survey from £15,890 to £13,450. This sum was allocated by the Treasury from the Development Fund.[11] Expenditure in the first year was actually less than anticipated, owing to the delay in the Advisory Centres obtaining the staff they required. However, some additional cost lay in the provision of filing cabinets, of which 100 had been ordered. These were to be distributed to Advisory Centres in proportion to the number of holdings for which they were responsible. Each filing cabinet could take the records of 3000 holdings.

Crop Reporters

These were Ministry of Agriculture and Fisheries employees (at least in part, since they may also have been employed by county councils) whose main task for the Ministry was to produce a monthly report on the crop prospects for the areas for which they were responsible, as well as supervising the quarterly agricultural returns for those areas.[12] Their other work appears to have been the gathering of statistics and estimates on the farm holdings that would be of use to agricultural planners, as well as advising on changes to holdings and their management and recording those alterations in the Parish Lists (for which, see later).

The 4 June Return (Form 398/SS) was returned by the farmer to the Crop Reporter whose address appears on the form, but the latters' principal role in the NFS was to clear up the errors arising from the inadequately completed page 3 Supplementary Form (see later). The areas for which Crop Reporters were responsible were numbered, and this number formed part of the farm code reference used by the Statistical Branch of the Ministry at St Anne's: thus in KT/46/128/9, the '46' is the Crop Reporter's number. The parishes making up each reporter's area can be reconstructed from the summaries of the Agricultural Returns, retained at the Public Record Office (PRO) as MAF 68. Some Crop Reporters worked on their own while others appear to have had staff of one or more individuals to assist.[13]

Land Commissioners

Appointed to advise on legal matters, these were senior local agents of the Ministry of Agriculture, indeed the 'eyes and ears' for the Ministry.[14] As offi-

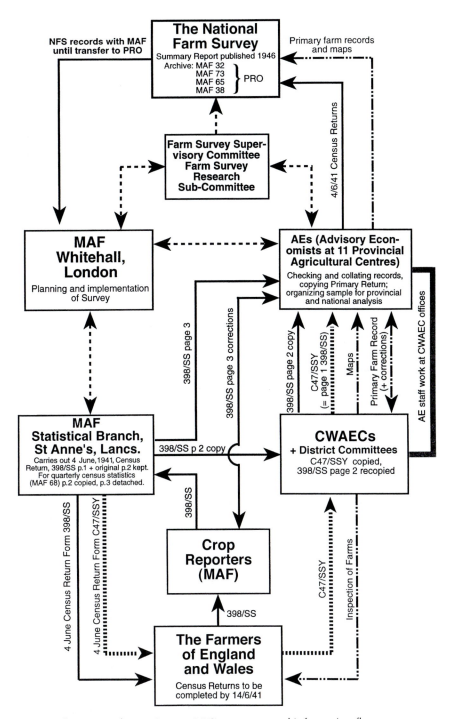

Fig. 3.4. The National Farm Survey (NFS): structure and information flows.

cials, the Land Commisssion (formally established by the Settled Lands Act 1882) had a long history stretching back to work with the Tithe Commissioners and the Enclosure Commissioners and with Copyhold abolition. The newly formed Board of Agriculture subsumed the Commission in 1889, but the Commissioners continued to carry out the board's responsibilities as professional or technical staff, especially working in the field (Parsons, 1969). During the war, the Commissioners soon became vital in such matters as the dispossession of farmers from parts or all of their holdings. Their remit might consist of one or more counties, for which they would also act in a liaison capacity with the Minister. Officials from the Land Commission service were subsequently enlisted into the Agricultural Land Service in 1948 (Foreman, 1989: 64).

The Farm Survey Supervisory Committee

This committee superseded the Farm Survey Committee. It was set up to give direction and coordination to the NFS, and consisted of 14 members, made up of MAF personnel, Advisory Economists and members of CWAECs. Its first meeting was held on 27 May 1941 (see Appendix).

The Farm Survey Research Sub-Committee

This was established in March 1942 to evaluate the farm records being obtained, to suggest how the information gained might best be used and to formulate a scheme for a national and provincial analysis. It consisted of six members – a mixture of MAF, CWAEC, and Advisory Centre personnel (see Appendix).

The Progress of the Survey

The various stages of the work for the NFS were coordinated into a complex web of interrelationships that involved the transfer of various elements of the survey forms around England and Wales (Figs. 3.4 and 3.5). The main stages can be outlined as follows:

1. Survey work on the farms by the CWAECs from 1941.
2. Copying and assembling of this survey work as the Primary Farm Return by the Advisory Economists' staff.
3. Receipt of the 4 June 1941 Census Returns at the Advisory Centres.
4. Matching of the two halves of the individual farm record (the Primary Return with the Census Return) at the Advisory Centre.

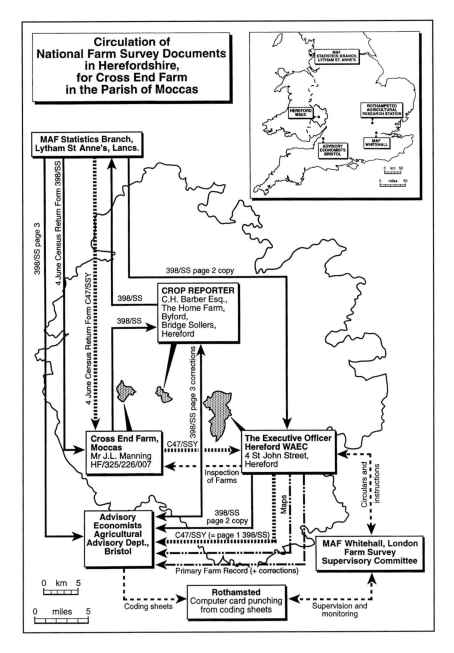

Fig. 3.5. The National Farm Survey (NFS): the farm/Ministry links as exemplified by Cross End Farm, Moccas, Herefordshire.

5. Receipt by the Advisory Economists of the farm boundary maps from the CWAECs (thereby completing the total record of the farms).
6. Selection of a sample of the records by each Advisory Centre to form the basis of a national statistical analysis of the farm survey data.
7. Analysis work completed by machine tabulation.
8. Publication by the MAF of a report in 1946 on the NFS and the analysis of its data.

A memorandum of 16 September 1940, initialled by M.G. Kendall at the Ministry, set out the aims and proposed composition of an extended farm survey, well before the initial survey was complete. Furthermore, on 14 November 1940 the Farm Survey Committee reported to the Ministry on the scope and content of the planned NFS.[15] As it was understood that further considerable survey fieldwork would be necessary, and that the completion of the necessary records would be laborious, an estimate of time for the survey of between 1 and 2 years was given. The Ministry was concerned that the CWAECs should now survey all farms, and not just the 'C' farms that had previously been the main focus of attention.

The exact purpose of the NFS was defined only later in the course of its slow progress towards completion. Indeed, a paper from the Farm Survey Supervisory Committee in February 1942, stated that: 'The objects of the farm survey, although well-known in a general way, have not been clearly defined.'[16] However, as part of a paper written in November 1941 entitled 'Notes on Publicity for Farm Survey', the purposes of the survey had actually been set out very clearly:

- *Wartime administration*: the survey would be used by the CWAECs to help to raise wartime farming standards.
- *Basis for advisory work*: the Advisory Services would be able to locate farms on which there was scope for further technical guidance.
- *Analysis*: a great amount of statistical analysis would be needed for post-war planning and administration. This would help with land management and improvement, and with the fuller utilization of scientific knowledge and more accurate criteria for economic efficiency.
- *Land planning*: what agricultural land should be reserved as such and protected from development for non-agricultural purposes.
- *Historical*: there would be a permanent record of the main features of every farm, comparable with the Domesday Book but more detailed and comprehensive. 'This record will be a mine of information for generations of historians.'[17]

In March 1942, Major Jeans, a member of the Supervisory Committee, wrote that:

> The most important and main object of the farm survey is probably to obtain the necessary information to enable a comparison of facts to be made, which

would assist the Government in taking immediate steps to increase the war production of food by cashing fertility without necessarily considering the economic financial result. This information and material, together with that obtained from quarterly agricultural returns, should provide facts for comparison and tabulation which would be of great value to the Government in its endeavours to evolve a post-war agricultural policy that would encourage greater efficiency and co-operation by permanently maintaining a prosperous and *self-supporting* agricultural industry.[18]

In 1943, the purposes of the NFS were further defined. Four were now stated:

1. To assist wartime administration by the CWAECs.
2. To assist in the protection of agricultural land from undesirable development and possibly to facilitate post-war regrouping and re-equipping of holdings.
3. To assist the advisory work of the agricultural colleges.
4. To contribute to the statistical and other information required for post-war planning and administration.[19]

The last point is specifically addressed by Yates (1945: 162), writing from Rothamsted: 'The farm survey will undoubtedly provide results which will serve as a basis for the permanent reconstruction of agriculture after the war.'[20]

The role of the Advisory Economists in the collation of the data from the survey was anticipated from November 1940, and the Ministry was at pains to inform the CWAECs that their work for the NFS would be limited to the actual survey work. The task of completing and assembling the records would be carried out by the Advisory Economists, who discussed the proposed new survey, and the necessary concomitant increase in their staffs, at meetings held on 6 November 1940, and on 10 January and 6 February 1941.[21] At the meeting on 6 November 1940, Professor Ashby said that he and others 'had been agitating for this sort of record for many years'. The proposal for the new extended survey was accepted at this meeting by the Advisory Economists in principle. At no stage in the planning of the NFS did the MAF submit the proposed survey to the War Cabinet for approval. Instead, decisions were made solely by the Ministry and the Advisory Economists, in collaboration with the CWAECs.

A further plan to supplement the new survey was set out in May 1941 by Dr F. Yates from the Rothamsted Experimental Station at Harpenden.[22] This plan allowed for the random sampling of some 6000–10,000 farms in order to gain technical and economic information that would supplement the 'broad general picture' of the NFS. Such a sample, he maintained, would also enable an earlier presentation of the results of the main survey. The plan was discussed at the first meeting of the Farm Survey Supervisory Committee held on 27 May 1941. Dr Yates emphasized that 'the additional work entailed by the proposal would be very small'. The idea was rejected, however, principally because of the extra burden that would undoubtedly have been placed on the CWAECs.

Then came Circular 545 dated 26 April 1941, from the Ministry to the CWAECs, which actually gave effect to the NFS. Information on each farm or other holding of 5 acres or more was now to be gathered in the form of a Primary Record, with a map showing its area and boundaries, and from its agricultural Census Return for 4 June, 1941. It was admitted that 'a certain amount of additional enquiry will be necessary', but the CWAECs were asked to complete their part of the survey by the end of March 1942. Unrealistically, on an earlier draft of the circular, the original date for completion had been stated as the end of September 1941.[23] A list of the Advisory Economists (with their addresses) who would be working on the survey in conjunction with the CWAECs was added to Circular 545 (see Appendix).

Further circulars from the Ministry now followed in quick succession. Circular 553, issued on 5 May 1941, congratulated the CWAECs on the work they had done on the first survey, and particularly emphasized the importance of the 'A, B, C' farm managerial grading and the new emphasis (i.e. by management ability rather than quality of farmland) of a revised classification under the new survey. On 31 May 1941 Circular 577 set out the proposed organization for the gathering and copying of the information from the 4 June 1941 Census Returns that was to form an important component of the NFS.

Circular 611 of 1 July 1941 alerted CWAECs that work on the new survey should begin immediately 'if it has not already been begun'. They were told that they could use the Primary Record form as a field sheet if they so required, and that copies of these had been issued to the Advisory Economists, to whom the CWAECs should apply. The first results of deliberations by the Farm Survey Supervisory Committee were included on the reverse of the circular, and the CWAECs were instructed to attach these rulings on how the Primary Return should be completed to their copies of the booklet, *Instructions for the Completion of the Primary Record*. The major points from these rulings were concerned with the 'A, B, C' classification: 'A' should mean that the farm was considered to be operating at 80% of its maximum production, 'B' at 60–80% and 'C' at less than 60%.

A further memorandum (No. 659) was issued on 8 August 1941 to Executive Officers of CWAECs. This set out the procedure for the custody of the 4 June returns, and the method by which the various parts would end up eventually with the Advisory Economists to form the total farm record. Attention was also drawn to the entry in the Primary Return of farmers' personal failings. Some CWAECs had told the Ministry that surveyors were reluctant to pass such judgement on farmers who might be their neighbours. The Ministry's view was that not all failings were culpable. 'Old age', 'lack of experience' or 'lack of ambition' should be entered as failings, but such judgements did not necessarily mean that the farmer was blameworthy. The surveyors were urged to be honest, and not to be concerned about the possibility of actions for libel, for there was no risk of this while they confined themselves strictly to their work of compiling the farm record. What they were not to do, however, and which could result in a charge of libel, was to spread information verbally about farmers to people who were not entitled to receive it. The memorandum confirmed that the CWAECs had powers under the Defence Regulations to empower farmers to give the answers needed for the Primary Return.

Yet another memorandum (No. 701) was issued on 17 September 1941. This was at pains to emphasize that the 'A, B, C' classification was to be based on the farmer's managerial abilities and not on the inherent quality of the farm. As the 1940 survey had combined managerial efficiency with the inherent farm quality, it was to be expected that gradings for the new survey would be different.[24] It was emphasized as well that surveyors should complete the entries on personal failings 'without fear or favour' to the farmer who was the subject of the report. Clarification was given regarding holdings such as prisons, or other institutions, or occupiers with very large gardens who were engaged in wartime food production on more than 5 acres but who would not do so in peacetime. Such holdings did not have to be included in the survey.

The outpouring of information to the CWAECs that issued from the Ministry in 1941 for the implementation of the NFS was eventually concluded for that year by the issue in November of the *Revised Instructions for Completion of Farm Records and Maps.*

The year 1942 was to be critical in the life of the NFS as the CWAECs had been asked to complete their role in the survey by the end of March of that year. However, in February there was still concern being expressed about the confidentiality of the managerial gradings. A memorandum from the Ministry to Land Commissioners stated that the managerial classifications should not be revealed to anyone other than the farmer themselves – not even to the landowners, who should be given information about their tenants in general terms rather than have the actual grading disclosed. The importance of working with landowners for increased food production was emphasized. The memorandum also reminded the CWAECs that they must maintain high standards in the grading of the farms. Whereas 'C' farms remained a top priority, the CWAECs must direct their energies as well to increasing the productivity of 'B' and even 'A' farms.

Also in February, a memorandum on the administration of the NFS was circulated to members of the Farm Survey Supervisory Committee for their meeting to be held on 5 March 1942. It had been written by J.H. Kirk and H. Whitby of the Ministry of Agriculture. This very comprehensive document provided the following summaries of information on the progress of the survey:

1. Cheshire had started the survey in the summer of 1941 but other counties, including the West Riding of Yorkshire, began only in the winter of 1941. The only county likely to meet the intended completion date of 31 March 1942 was Cheshire. However, it was estimated that other counties would complete between April and September 1942. Delays had been caused not only by the greater time needed than that estimated for the completion of each farm record, but also by the fact that there had been a shortage of staff (paid and unpaid) employed on the survey by the CWAECs and the Advisory Centres.
2. Some counties would complete late not because of the type of difficulties set out above, but because they had set themselves particularly high standards. Cornwall and Gloucestershire were two such counties.
3. There had been several references in the press to the survey, and also in Parliament. It was proposed to arrange for some systematic publicity in the near future which would be of general interest to the public and would help to stimulate the efforts of survey workers.
4. It was known that enthusiasm for the survey was not high amongst CWAEC district committees and other fieldworkers. The CWAECs themselves had shown little enthusiasm, but this was changing as the uses of the survey became clearer. The Land Commissioners and Advisory Economists had been the most keen to see the survey developed to the full. The Economists were thanked in particular for their unflagging efforts, despite having two to three times the expected work unloaded on them.[25]

A further document distributed to the Farm Survey Supervisory Committee for the meeting on 5 March suggested that there should be additional 'B+' and 'B−' managerial ability classifications, as otherwise 'B' covered a very broad range of difference from 'bordering on A' to 'bordering on C'. This suggestion was later accepted, but a further proposal that both the farmer and the farm should be classified was not taken up.[26]

The last document presented to the Farm Survey Supervisory Committee was a paper written by Kirk that restated the aims of the survey and the future use of the information gathered. Consideration was also given by him to the permanent preservation of the records of the survey (as a 'Domesday record'). Storage space, it was thought, would be a problem, as would be the need for everything that had been entered 'in pencil or crayon' to be 'done over in ink or paint'. Kirk's conclusion was that 'it will probably be desirable to seek the advice of a professional archivist' for papers other than those relating to the final analysis of the data.

The meeting on 5 March 1942 was indeed therefore an important one for the progress of the NFS. The major subjects that were actually discussed by the Supervisory Committee, and their conclusions and recommendations, can be summarized as follows:

1. There was concern about the differing standards of the 'A, B, C' grading. One entire district of Devonshire had a 98% 'A' grading. It was felt that one of the reasons for this was that the farm surveyors had no idea of national standards, or even of the overall standards within their own county. A proposal that the CWAECs be sent a statement showing the farm gradings that had been received so far for the whole country was rejected, as it was thought that this would add to the pressures on the surveyors not to return an accurate grading. It was agreed that, whereas no overall change in the 'A, B, C' classifications should be made (these would be used for the planned national analysis), individual CWAECs could, if they felt this would help them, grade by five categories: 'A', 'B+', 'B', 'B−' and 'C'.

2. The question of rough grazing was discussed, and more particularly the lack of provision in the Primary Return for a proper description of this. The CWAECs for Durham, Northumberland, Cumberland and Westmorland had agreed to supply, in the General Comments section of the Primary Return, extra details on rough grazings, including areas of 'white' and 'black' grazing, bracken infestation, heather (whether burnt or not), drainage and erosion. However, it was felt that CWAECs should only supply information of this sort on rough grazing where it was considered that this was of particular importance. In most lowland counties, it was thought, much of the rough grazing was not important enough to merit extra description.

3. Consideration was given to using the Farm Records as a survey of estates. It was thought that this work would generate too much detail for a general analysis to be made subsequently, but that it might be possible to look at the estates of individually selected owners.

4. The question of dispossessions resulting from bad husbandry was discussed. It was felt that one of the reasons for the need for wartime dispossessions lay in the Agricultural Holdings Act 1922, by which landlords could not eject tenants unless they obtained first from the county council Agricultural Committee a bad farming certificate – this was notoriously difficult to obtain (there had been only three cases in Wiltshire in 15 years). A committee member wished the CWAECs to supply lists of dispossessed farmers, and the chairman stated that, although this was beyond the scope of the present survey, he would approach the appropriate Ministry department on the matter.

Following this meeting of the Farm Survey Supervisory Committee on 5 March, Memorandum 858 of 23 March put into effect much of what had been discussed at the meeting, such as the new managerial grading categories that CWAECs were now allowed to apply to the Primary Return, and the rough grazings descriptions. In addition, the CWAECs were told that they could, if they wished, mark the condition of arable land and of pasture by percentages spread over 'Good', 'Fair', 'Poor' and 'Bad', but that, if they did this, the principal category should also remain, indicated by a cross.

Further detail for the completion of the survey returns was discussed in a memorandum of 13 May 1942 from the Ministry to Advisory Economists. This largely concerned the machinery and labour that was drawn from the CWAEC pool or from private contractors, but which was present on individual holdings at the time of the Census Return of 4 June 1941. Most farmers were excluding details of the machinery and labour from these sources, whereas it was the Ministry's intention that they be included. The Ministry sought the views of the Economists as to whether this data should be included or excluded from the national analysis. Other points from the memorandum concerned the names of parishes that did not match with the Ministry's code reference, the definition of the area of a farm and the rental value of holdings principally under glass.

The original date of 31 March 1942 for the completion of the CWAEC role in the NFS having long since passed, the Ministry now selected a date 6 months further on (30 September) as that by which to judge the overall progress of the survey. As that latter date approached, a paper reviewing the current position was produced by the Ministry, although much of the information in this review related to the situation as it was in mid-July, with progress by the CWAECs towards completion merely anticipated up to September and December. The details were as set out in Box 3.1. These lists excluded Welsh counties, for which details were not available but which were said to fall within groups (iii) and (iv) rather than (i) and (ii).

The only counties that had made any real progress with the next stage of matching the Primary Returns with the Census Returns were Kent and the North Riding of Yorkshire, although it was hoped rather vaguely that most counties had at least made a start with this work. Certainly by the end of 1942 one CWAEC chairman, Lord Cranworth from East Suffolk, was noting that no guidance 'from above' had been received regarding a yardstick for the classification of the county's farms as 'A', 'B' or 'C'. In his own district committee they had therefore evolved their own formula, with 'A' farms producing at least 75% of maximum productivity, 'B' between 50% and 75%, and C below 50% (Burrell *et al.*, 1947: 82). The situation with the maps that were to form a vital part of the complete farm record was even less advanced, partly because the CWAECs had been told to leave this task to the end.

The final two stages in the survey – the selection of a national sample of the records, and the annotation of that sample so that it could be entered on punched cards for analysis – still lay in the future. Once successful pilot samples had been selected and analysed, it was hoped that a start on the work for the full national analysis might be made by February 1943.

At the conclusion of 1942, the second year of the NFS, a press release was issued on 17 December in which the wartime survey was compared with the Domesday Book.[27] One of the findings of the survey, it was said, was that the nation's agricultural land was highly fragmented. Thousands of

Box 3.1. The progress of the Survey by mid-1942.

(i)   *Counties which would have completed the fieldwork, the copying of the Primary Record and the checking of the Census Returns by 30 September 1942:*
Northumberland, Yorkshire North Riding, Yorkshire East Riding, Essex, West Suffolk, Nottinghamshire, Herefordshire, Bedfordshire, Cambridgeshire, Norfolk, Kent, Surrey, West Sussex

(ii) *Counties which would have completed the fieldwork only by 30 September 1942:*
Cheshire, Staffordshire, Rutland, Lincolnshire (Holland), East Suffolk, Soke of Peterborough, Berkshire, Buckinghamshire, Dorset, Hampshire, Middlesex, Isle of Wight, Northamptonshire, Devon, Cornwall, Westmorland

(iii) *Counties which would have completed the fieldwork by 31 December 1942:*
Cumberland, Durham, Yorkshire West Riding, Lancashire, Shropshire, Warwickshire, Derbyshire, Leicestershire, Lincolnshire (Lindsey), Lincolnshire (Kesteven), Hertfordshire, Isle of Ely, Oxfordshire, Gloucestershire, Wiltshire, East Sussex

(iv) *Counties which would <u>not</u> have completed the fieldwork by 31 December 1942:*
Huntingdonshire, Worcestershire, Somerset

Source: PRO MAF 38/212 (12); 209 (FSC 14); 38/207 (51)

so-called farms existed merely as parcels of land strung out amongst several parishes, and these did not make economic farming units.

The new year was well advanced before a further memorandum (No. 1267) concerning the survey was issued by the Ministry. Dated 18 May 1943, this instructed the CWAECs not to disclose NFS information to anyone other than the Ministry's officers, including Rural Land Utilization Officers, as any disclosure might constitute a breach of Defence Regulation 84. This instruction was also made to the Advisory Economists.

The next major event in the history of the NFS was the meeting of the Farm Survey Supervisory Committee on 26 July 1943, for which a Farm Survey Progress Report was prepared by Kirk and Whitby.[28] This stated that the fieldwork for the Primary Returns was now 95% complete, and for the maps about 70% complete. The 5% uncompleted fieldwork was said to be concentrated in a few counties only. However, most counties still had queries to resolve, even though it was now 16 months after the date by which the CWAECs' role in the survey should have been completed. The report drew attention to four problems identified in the completion of the Primary Return:

1. The question 'Does Farmer Occupy Other Land?' was meant to apply where the other land was the subject of a separate Primary Return, and not where it was part of an amalgamated holding (for which there should be just one Primary Return). Advisory Economists were asked to amend the records selected for the national sample, if necessary, to ensure there was conformity in respect of this.

2. On the difficult question of the definition of a distinct farm unit, the Ministry set out the questions that should be asked:

(a) Were the holdings under the same detailed day-to-day management?

(b) Were they at a small or great distance apart?

(c) Did they have common, or separate, supplies of labour, machinery, etc. and, if the latter, what was the degree of mobility of these supplies among the holdings?

(d) Were the holdings, considered individually, capable with their present equipment, cropping and stocking and at contemporary price levels of being satisfactory economic units – in other words, could they have provided a sufficient living if farmed independently along present lines?

Box 3.2. Research at Advisory Centres related to the National Farm Survey (NFS).

Aberystwyth
'the preliminary work of completing the records for all farms, sampled and non-sampled, and the assembling of dossiers is expected to be finished during the first half of the year. Later on, it is hoped to add a copy of the 4 June return for either 1944 or 1945 to each of the existing records'.

Seale Hayne
'it is expected that the preliminary work of completing and checking the records and returns, the making up of dossiers etc, will take a further 12 months'.

Manchester
'the survey records and those obtained by other advisers, particularly the Chemist and the Entomologist, will be 'married''.

Wye
'the intention is first of all to complete the Farm Survey dossiers, which will include material provided by other Advisers (eg the Advisory Chemists) at this Centre, and to prepare a card index of farms ... '.

3. The question 'Class of Farmer': a further class of farmers was suggested, 'Professional and Business'. The purpose of this was to gain a category for farmers who otherwise followed professional careers (teachers, company directors, solicitors, etc.) and for whom the part-time and spare-time categories were not really applicable.

4. The question 'Water Supply': the type of water supply should reflect the source and not the conveyance, i.e. water piped from a well should be placed under 'Well' and not 'Pipe'.

In May 1944, a statement that set out the programmes of research on the NFS material to be carried out at the Advisory Centres included the points on the position with the farm records at that date (Box 3.2).

The National Farm Survey Documents

Having examined the organizational structure of the survey, and its progress, we now describe in some detail the main elements of the survey in turn, as they now exist at the PRO.

4 June 1941 census forms

The blank census forms for the 4 June 1941 agricultural return were sent by the Statistical Branch of the MAF at St Anne's-on-Sea, Lancashire, directly to the farmer, who was instructed to complete both forms by 14 June and return them to the addresses printed on the forms. The mailing lists for the quarterly Census Returns were maintained as an addressograph system and it is these records that are known as the Parish Lists. From 1939, quarterly returns of a more limited nature than the June returns were also collected in September, December and March. In 1953 these were replaced by statistical sampling, but no parish returns of these summaries appear to be available (Coppock, 1955: 13).[29]

1A. C47/SSY (BLUE FORM)
Single-sided form with questions 1–89, covering the sections 'Crops and Grass', 'Livestock', and 'Labour', to be completed by the farmer and sent direct to the CWAEC. The address was on the reverse of the form, which was folded so that it could be sent through the post. The forms are dated either April 1941 or June 1941.[30]

1B. C49/SSY
This is the version of C47/SSY used in Wales. The form is in Welsh as well as English, and, because of this, is slightly larger than C47/SSY.

1C. 398/SS (GREEN FORM) (see Fig. 1.2)

A three-page form with questions 1–148. Page 1 contains questions 1–80, but set out in a different way from C47/SSY. In 398/SS the questions on 'Labour' (35–42 on C47/SSY) become questions 73–80, and the 'Livestock' questions covering cattle, sheep, pigs, poultry and goats become questions 35–72 (instead of 43–80 in C47/SSY). 'Horses' (questions 81–89 in C47/SSY) becomes questions 116–124 of page 2 of 398/SS. 'Small Fruit and Vegetables' forms questions 81–115 of page 2 and 'Stocks of Hay and Straw' 125–126. Questions 127–128 of page 2 relate to 'Changes of Occupation of Land between 1940 and 1941'.[31] On page 2 the Occupier signed the return and completed his/her postal address. Page 3 (which is marked 'S.F.', for Supplementary Form) contains questions 129–148 covering 'Labour' (supplementary questions), 'Motive Power', 'Rent', and 'Length of Occupation'. These page 3 questions were included especially for the purposes of the farm survey; they did not form part of the usual Census Returns. The three pages of form 398/SS were returned by the farmer to the address of the Ministry's local Crop Reporter that was printed on the back page, the form being folded so that it could be sent through the post. As with C47/SSY, there is a Welsh version of this form, but its reference number has not been identified.[32]

1D. C51/SSY

This was the form on which details of 'Small Fruit and Vegetables', and 'Stocks of Hay and Straw' (questions 81–115 and 125–126) were copied by the Statistical Branch from Form 398/SS. It was then sent to the CWAECs, who prepared a further copy for their purposes (also on C51/SSY) and sent the original copy to the Advisory Centre. The form is dated May 1941.[33]

1E. C69/SSY

This was the beige-coloured (termed 'white') reminder form dated July 1941 that was sent to farmers who had failed to answer the supplementary questions on page 3 of form 398/SS (Fig. 1.4). It was often printed with the return address of the Ministry's Statistical Branch at St Anne's, and again it could be folded to be sent through the post.[34] In some instances, a version of the 398/SS page 3 supplementary form, yellow/green in colour and on thicker paper, was used as the reminder form. C69/SSY was sent out with a covering printed note (M19524/5644) drawing the farmer's attention to the Defence Regulations 1939, and the Agricultural Returns Order 1939, under which the information was compulsorily demanded, and instructing that the questions marked with a red 'X' on the C69/SSY must be answered (Fig. 3.6).[35]

The Primary Return (Fig. 1.1)

The data forming the Primary Return was obtained in the field by the CWAECs and sent by them to the Advisory Economists, where the information was often copied. The Farm Survey form was given the reference number B496/EI. It was sometimes made up in pads of 50, with a stiff back to each pad, to be used in the field by the CWAECs.[36] Generally, however, the version of the form used by the CWAECs, or by the Advisory Economists' staff to copy the information supplied by the CWAECs, seems to have been supplied as loose sheets. The first printing of the form came in 3000 pads in March 1941, but further printings took place in June, July and October 1941. A smaller version of the form, with the same complete questions but with the farm identity details compressed in a box in the top right corner of the form, was issued in August 1942 as 520 pads.[37]

(A)

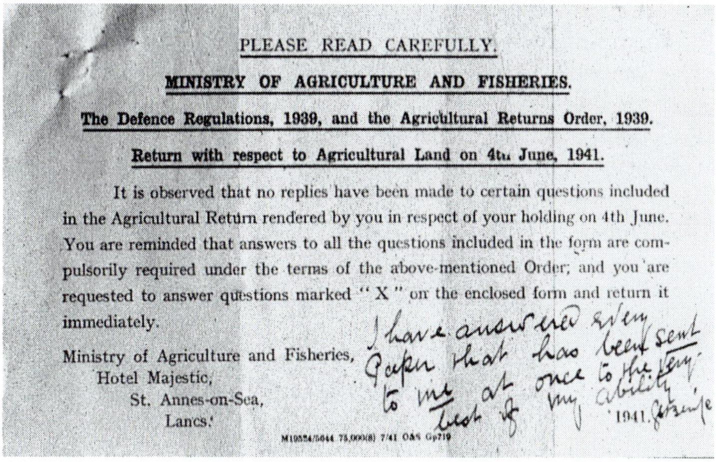

PLEASE READ CAREFULLY.

MINISTRY OF AGRICULTURE AND FISHERIES.

The Defence Regulations, 1939, and the Agricultural Returns Order, 1939.

Return with respect to Agricultural Land on 4th June, 1941.

It is observed that no replies have been made to certain questions included in the Agricultural Return rendered by you in respect of your holding on 4th June. You are reminded that answers to all the questions included in the form are compulsorily required under the terms of the above-mentioned Order; and you are requested to answer questions marked "X" on the enclosed form and return it immediately.

Ministry of Agriculture and Fisheries,
Hotel Majestic,
St. Annes-on-Sea,
Lancs.

I have answered every paper that has been sent to me at once to the very best of my ability

1941.

M19354/5044 75,000(8) 7/41 O&S Gp719

(B)

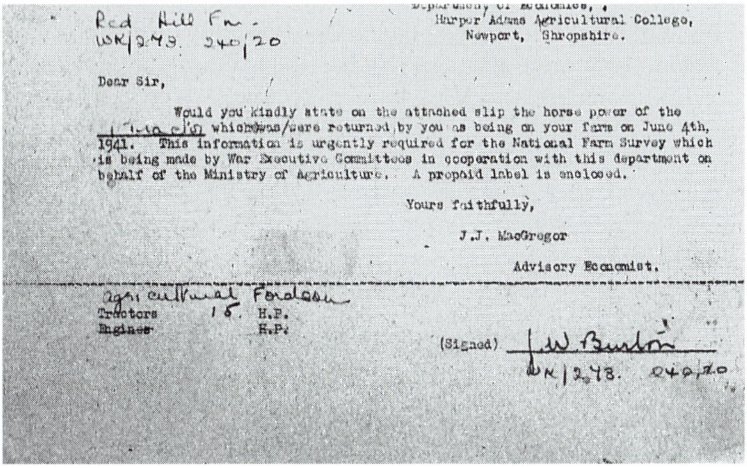

Red Hill Fm.
WK/2.43. 240/20

Department of Economics,
Harper Adams Agricultural College,
Newport, Shropshire.

Dear Sir,

Would you kindly state on the attached slip the horse power of the tractor which was/were returned by you as being on your farm on June 4th, 1941. This information is urgently required for the National Farm Survey which is being made by War Executive Committees in cooperation with this department on behalf of the Ministry of Agriculture. A prepaid label is enclosed.

Yours faithfully,

J.J. MacGregor

Advisory Economist.

agricultural Fordson
Tractors 15. H.P.
Engines H.P.

(Signed) J.W. Burton
WK/2.43. 240/20

Fig. 3.6. Two reminder notices issued in 1941 for (A) the 4 June Returns for that year (with a weary response added), and (B) the National Farm Survey. (Source: PRO MAF 32/366/112 and MAF 32/964/240, is Crown copyright and is reproduced with the permission of the Controller of Her Majesty's Stationery Office.)

As we have already seen, the Farm Survey Primary Record form was developed out of recommendations by the Farm Survey Committee in October 1940, and by a sub-committee of the Conference of Advisory Economists during the same month (see Appendix). The Advisory Economists wanted a much more detailed form than the one eventually accepted by the Ministry, and there is evidence that they were not satisfied with the final result. There were original proposals for questions on, for example, the highest and lowest altitude of the holding, whether the fields were in one block or several, the types of field boundaries, the actual distance of the farmstead from the nearest main road and railway station, and the main products sold from the farm. A typed early version of the questions that the Ministry agreed should be asked included those on the types of drainage of fields, on the taking of soil samples, on the farmer's keeping of accounts, on the type of slope on which the farm was situated and whether it was in an exposed or sheltered position. An amended version of these questions, deleting several of the original suggestions, was drawn up in a typed form known as the Farm Survey Record Form, and a later version of this was produced, with certain alterations.[38]

In November, the form was revised again to include questions on rent and length of occupancy.[39] These questions were later deleted from the Primary Record form and added to page 3 of the 398/SS Census Return – thereby creating one of the major problems that prevented an easier and earlier completion of the NFS. A manuscript single-sided sheet setting out the questions, with further alterations, in the style that was eventually adopted for the B496/EI form is undated, but is with correspondence of December 1940 to January 1941.[40] There was a first typed version of the B496/EI form before the final printed version appeared in March 1941, incorporating a number of small alterations.[41] Circular 594 to the CWAECs dated 13 June 1941 drew attention to these alterations, which concerned the form of questions B1 and B3 where percentages were now required. Memorandum No. 659 of 8 August 1941 instructed surveyors to add the acreage figure of the land to which the survey related in the top right corner of the record, above the code reference. This was particularly important as it would help in the eventual matching of the Primary Returns with the Census Returns.[42]

A typewritten draft of the *Instructions for the Completion of the Primary Record*, to be used by the surveyors from the CWAECs, appeared in about December 1940. Greatly expanded versions of the instructions were drawn up later, and in May 1941 they were assembled in a small buff-coloured booklet. *Revised Instructions for Completion of Farm Records and Maps* (in a similar format but now on green paper) were issued in November 1941. The instructions were regularly updated through the medium of circulars and memoranda to the CWAECs throughout the period of the farm survey work.[43]

The Maps

As part of the complete individual farm record, the CWAECs were instructed to delineate the boundaries of each farm on a set of 6-inch Ordnance Survey sheets.[44] The necessary map sheets were to be supplied by the Ministry of Agriculture to the CWAECs, if required. In the Ministry's Circular 545 to the CWAECs (26 April 1941) it was stated that, as the work would undoubtedly place a considerable extra burden on the CWAECs, it was not expected that the mapping work would be completed with the Primary Return by the end of March 1942. The Committees were merely asked to proceed with the work as quickly as their other duties would allow.

A Ministry memorandum of 21 March 1941 had already made clear the importance of the maps to the survey, and stated that the inclusion of a map, as part of the total farm record, meant that questions on such matters as altitude, slope, farm roads, etc., as part of the Primary Return could thus be omitted.[45] However, the feeling at the Ministry was that this part of the survey could be left 'for the present'. The Advisory Economists had also expressed reservations about the maps, mainly on the grounds of the difficulties of storing them.

Circular 611 of 1 July 1941 from the Ministry to the CWAECs suggested that the Committees might consider approaching institutions and authorities within their areas to see if these bodies could lend maps for copying. County councils and private estates might well have maps that could be loaned, and it was possible that farm boundaries might also be determined from the maps in the possession of the District Valuer.

Memorandum 701 of 17 September 1941 informed CWAECs that where 6-inch sheets were used for the NFS mapping, the Ordnance Survey parcel (or field) numbers must be transcribed from 25-inch sheets on to them. The memorandum confirmed that 6-inch sheets should be used generally, but, where these were not available as a result of losses by bombing or for other reasons, then the CWAECs should use 25-inch sheets, photographically reduced to a $12\frac{1}{2}$-inch scale. The CWAECs were asked to send clean copies of 25-inch sheets (rolled, not folded) to the Ministry's offices at Soho Square, London, for the photographic reduction work to be done. Three bromide prints at the reduced scale would be returned with the original map sheet. In fast time 20,000 25-inch sheets were photographically reduced in this manner, 7–10 days being taken for the work to be done.[46]

A proposal made in October 1941 to show the Ordnance Survey parcel numbers on the Primary Record rather than on the 6-inch map sheets was quietly dropped, and it was confirmed that the numbers must be transcribed from 25-inch sheets on to the 6-inch sheets. In November 1941, the CWAECs were informed that they could use the 25-inch reduced maps to show small holdings whether or not 6-inch sheets were available. The 25-inch sheets proved to be more popular with CWAECs, as they obviated the need for the labour-intensive transcription of the Ordnance Survey parcel numbers (Fig. 1.5).

A meeting of the Farm Survey Supervisory Committee on 27 May 1941 recommended that coloured pencils be used to show the farm boundaries.[47] Instructions on the completion of the maps were omitted from the first issue of the booklet of *Instructions for the Completion of the Primary Record*, but the issue of the *Revised Instructions* in November 1941 included those for the maps. Under these instructions, the CWAECs were required to use 6-inch or 25-inch (reduced) sheets on which the farm boundaries should be delineated in ink. The name of the occupier and the name of the holding (or the farm code number) were to be shown, together with the Ordnance Survey parcel numbers where necessary, although the acreage figures were *not* required. Multiple holdings were to be linked by a system of bracketing or colouring, and if there were no multiple holdings, there should be no need for colouring. Finally, the separate parts of a holding were to be cross-indexed where they appeared on adjacent map sheets.

In August 1941, the Ministry informed the CWAECs that, for security reasons, they were not to show aerodromes, camps, battle areas and other military land on the maps. In March 1943, when the matter was raised again, the suggestion was made that military land should be marked 'not used for agriculture' or that the areas in question should be marked with a dummy code number. However, the Advisory Economists pointed out in a memorandum to the Ministry that their staff were still finding maps showing 'aerodromes' and 'RAF ammunition dumps'![48]

Instructions issued in September 1943 said that each map sheet that had not yet been completed should be dated. The date required by the Ministry was that on which the main data on the farm boundaries had been determined. The reason for the dating was to emphasize the probable date gap between the survey for the Primary Record and the drawing up of the map. There was no requirement for the CWAECs to date maps that had already been completed, but it was hoped that a rough date (e.g. 'September 1942' or 'Spring 1943') might be added.

The survey process

The completion of the Primary Return

The role of the CWAECs in the NFS began in May 1941. All farms and other agricultural holdings of 5 acres and more were to be surveyed, and the district committees of the CWAECs were responsible for the survey of the farms in their areas. It was hoped that the information required on certain farms would be already available, or could be obtained in the course of visits to farms in connection with the main purpose of the Committees – that of increased food production. Special visits to farms for the purposes of the survey were to be avoided if possible, but the reality of the survey, in fact, demanded that many be made. Most of the information on farms already available to CWAECs related to the poorer farms (the 'B' and 'C' farms) that needed assistance and supervision to gain greater productivity. The NFS now demanded that all holdings, whatever the standard of their condition and management, be inspected, and much of this could only be done by considerable additional fieldwork. In Montgomery an attempt to gain the requisite information was made through a postal questionnaire, but this was later frowned upon by the Farm Survey Supervisory Committee at their meeting on 5 March 1942. The Ministry of Agriculture had only agreed reluctantly to this method of completing the Primary Return and had advocated safeguards (not stated) 'against the more obvious weaknesses'. An example of the process operating at the local level and its interface with the national level is given for Cross End Farm, Moccas (Herefordshire) (see Fig. 3.5).

The press release on the NFS issued on 17 December 1942 notes that the usual procedure was for the surveyor to visit the farm, make a preliminary general inspection of the crops, livestock, buildings, etc. and then find the answers to the specific questions asked for in the Primary Return.[49] On the more factual questions, the farmer was asked directly for the information, but where a judgement was required, the surveyor had to form his own impression, without being swayed by the farmer concerned. The surveyor then returned to the district committee's office with a sheaf of notes and a rough plan of the farm, and these in due course were transcribed on to the farm record form and the map.[50] The original intention had been for the survey work (what the Ministry termed 'the collection of information') to be completed by the end of September 1941, but this date was later revised to 31 March 1942, and then to 30 September 1942.

The CWAECs had the power under the Defence Regulations to require farmers to furnish the necessary information, provided it was required for war purposes. The Ministry stated that all the questions on the Primary Return were asked with war purposes in mind, with the possible exception of those concerned with water and electricity, which might only be used to help to frame post-war policy.

Where farms had already been inspected and the required information collated according to whatever system the district committee used, the Recorder from the responsible Advisory Centre who would be working at the district committee's offices would copy this on to the Primary Record form (B496/EI). Where inspections were to be carried out by district officers, the salient information could be recorded in the field using the committee's own forms, or in note form, or by using the Primary Record form as a field form. In the latter case, this could then be recopied by the Recorder in the office, or it might be full and neat enough to stand as a completed record without the need for recopying. The Recorder's work did not consist simply of copying: each Primary Record would be checked for completeness and internal consistency, and unsatisfactory records would be referred back to the CWAECs for further attention. In some cases, this meant a partial re-survey of the holding in question.

Shropshire CWAEC issued instructions to its district committees concerning tasks for the 1941 season, including the NFS. Two copies of their form 227B/42 were issued for each farm. The notes stated that the 'Area Economist' would 'have access to the office files to extract all the information available' in order to enter up the Primary Return. They also said that 'the National Farm Survey must not be allowed to hold up your returns for land to be ploughed, drained etc and this would occur if an attempt was made to get all the information at once'. The additional information for the farm survey was then itemized – for example: 'What fire precautions in buildings and fields?' Shropshire also had a system of farm classification that ran A1, A2, A3, B1, B2, B3, C1, C2, C3, where the letter referred to the standard of management and production, and the number to the inherent quality of the soil. For example, A3 was 'a good farmer, maximum production, but bad land' and C1 was 'a bad farmer, bad production and good land'.[51]

The signatures on the Primary Record form should relate to the following:

(a) 'Field Information recorded by' = The CWAEC surveyor.
(b) 'Primary Record completed by' = The recorder from the Advisory Centre who has copied the record (but see (c)).
(c) A third name (or initials) = The recorder from the Advisory Centre who has copied the record. (If there is such a third signature, then signature (b) is likely to be that of a second CWAEC official who has checked the record.)

The surveyor's name would only be his actual signature if the Primary Return was completed by him and not recopied by the Recorder, whose own actual signature or initials should appear. Some Primary Records show names (a) and (b) as the same – presumably in these cases the record was made and completed by the surveyor and not recopied by the Recorder. Sometimes, too, one name (or both) has been stamped on the record, using an inked rubber stamp.

Completed Primary Records would presumably be taken (or possibly mailed) to the Advisory Centres by the Recorders. Although there is no direct evidence, it can also be assumed that the CWAEC district office kept a copy of each Primary Record for its own files.

Any problems or queries in regard to the Primary Record would be referred by the Recorder to the Advisory Centre supervisor, who would visit the CWAEC district offices on a regular basis to coordinate the work of the survey. It is also possible that the supervisor did some of the copying work, but this is not certain. Many of the Primary Records show signs of amendment and alteration made by someone checking the record, and presumably this was the work of a supervisor

It was decided late in June 1941 that the NFS should use for its farm identification reference the same coding system as was used by the Statistical Branch of the MAF at St Anne's. This code was defined as in the following example:

KT/	121/	283/	15
County (Kent)	Crop reporter's area	Parish	Schedule (farm)

The reference 'KT/121/283/15' would thus identify uniquely the particular farm. For convenience, within a local area, often the abbreviated form '283/15' would be used. The code could be further extended to 'KT/121/283/15/163.5/2', which adds to the reference the facts that the holding consists of 163.5 acres of Crops and Grass and 2 acres of Rough Grazing. However, it is relatively unusual to show the acreage figures attached to the farm identifier code in this way. Invariably, on the Census Return forms, these two elements of the Ministry's coding system are separated.

The coding system appears to have created some confusion with the district committees, who seem to have had their own county coding systems. Most confusion came from the identification of the district. The CWAECs' system of administrative districts followed that of the county council Rural Districts, which were usually given a code number by the CWAECs. This system was at variance with the Crop Reporters' areas that formed part of the Ministry's code, and sometimes an 'incorrect number' will appear for that of the Crop Reporter, which usually means a confusion between the CWAEC district number and that of the Crop Reporter's area. Very often both the county code and that of the Ministry are given on the Primary Return, and sometimes the former will be shown most prominently, with the Ministry code added above in pencil. On the maps, there are many cases of the county code system being used exclusively and the Ministry code not appearing at all.

When it was decided to use the Statistical Branch's system of farm identification codes, there were various proposals as to how the Advisory Economists could obtain details of these codes. One proposal was to copy the Parish Lists; another was that all the Primary Records should be sent to the Ministry at Whitehall, who would fill in the appropriate code references; a third was that Advisory Economists' staff should visit St Anne's and refer to the Parish Lists kept there. An official at St Anne's then pointed out to the Ministry planners that as the Advisory Centres would be receiving (either in original or in copy) the various forms making up the 4 June 1941 Census

Return – which included the code reference – the problem could be solved in that way. There is no indication on the files as to how exactly this worked. It must have involved an immense amount of time and labour checking completed Primary Records against Census Return forms in order to find the required code number. Was this done at the CWAEC district offices or back at the Advisory Centre, or was the code obtained from the copies of the census forms that were sent to the CWAECs? In most circumstances the code appears to have been written on the Primary Record at the same time as it was copied, though at times it does appear to have been written in another hand and at a later time. It is likely that in many cases the code number was not added until the matching process was carried out at the Advisory Centre.

A memorandum of 7 July 1941 indicates that there was a considerable delay (possibly until the end of September 1941) before the CWAECs could release the C47/SSY form to the Advisory Economists' staff for the purposes of gaining the code references, and that similarly, with the page 2 and page 3 form to be copied and detached from the 398/SS main return, there would be a lengthy delay before these would be available to the Advisory Economists.[52] It is not clear how the problem was eventually solved. It is possible that staff of some of the Advisory Centres did visit St Anne's to copy code references from the Parish Lists, but it seems most likely that the codes were obtained by the Recorders working at the CWAEC district offices, who had an early sight of the C47/SSY form. This would have meant working hand-in-glove with the CWAEC staff, and it is a compliment to the diplomacy that must have been exercised by the Recorders and supervisors from the Advisory Centres that there is no record of any friction between the various staff in this demanding and painstaking work.

Sometimes in a parish there are several Primary Records that have been given no code at all. This can occasionally happen when the holding is part of an amalgamated holding, perhaps with the main part of the holding in another parish from which the Census Return has been made. Sometimes, in cases like this, more than one Primary Record is made, and often the code reference will not be known. In other cases it has to be assumed that either the code reference could not be found, or perhaps there was no Census Return made – this can sometimes happen, for example, in the case of private houses with large grounds, part of which had been used for agriculture since the start of the war. If no Census Return had been asked for, there will be no code.

In order to complete the Primary Record, according to whatever system was actually used, the Recorder would copy in the farm's individual code reference. However, in addition to the Advisory Economists, the CWAEC district offices also needed full records of the code references for the holdings in their particular areas. The CWAECs must have obtained these either from a copy of the schedules being used by the Recorders, or by keeping copies of the Primary Returns, or from their own copies received of various pages of the 4 June Census Returns. The code references were needed, in particular, for transference to the maps on which each farm should have been indicated by its code. Some CWAECs, however, did not show the Ministry's code references on the maps at all, but used their own coding system to identify the holdings. This would seem to indicate either that these counties had no readily available record of the Ministry's codes, or that the importance of using those codes had not been sufficiently impressed upon them.

In addition to the farm code reference, another piece of information that often appears to be added at a later date to the Primary Return is the code of the map sheet reference to the location of the farm. This was evidently information that the surveyor did not always add to his survey notes, and it had to be found separately, either at the district committee office or at the Advisory Centre. In order not to hold up the despatch of records from the CWAECs to the Advisory Centres (the Wye Advisory Centre, for example, was completing 1000 Primary Returns a month, including filing and indexing) it was agreed that the number and edition of the relevant Ordnance Survey sheets could be copied on to the Primary Record 'at a later stage'.

Completion of the 4 June 1941 Census Return

The two census forms, C47/SSY and 398/SS, were mailed on 3 June 1941 from the Statistical Branch of the MAF at St Anne's (they appear to have been bulk posted from Manchester) direct to the occupiers of the farms and other holdings. The accompanying *Instructions for Completing the Return* (form SSY 2553) is also dated 3 June 1941. The farmers had until 14 June to complete the two forms and return the C47/SSY to their CWAEC and the 398/SS to their local Crop Reporter.

On receipt by the CWAECs, form C47/SSY was copied, using blank copies of the form supplied by the Ministry. CWAECs needed the information on the form in particular to administer the rationing of animal feedstuffs. The copy (or, more often, the original) was then sent to the appropriate Advisory Centre. The Ministry asked that this be done by mid-November 1941. On receipt by the Advisory Centres, the forms were checked and filed away to make up part of the complete farm record for each holding.

On receipt by Crop Reporters, forms 398/SS were checked and collated by parish for the Crop Reporter's area, and forwarded to the Statistical Branch at St Anne's. Here, the forms were checked to see if they had been completed fully and correctly, and pages 1 and 2 were tabulated to gain the information required by Statistical Branch for their regular quarterly return statistics. A copy was made on Form C51/SSY of the information on page 2 of the 398/SS form (i.e. the information on 'Small Fruit, Vegetables and

Flowers', and 'Stocks of Hay and Straw', but excluding the information on 'Horses' that was already available on the C47/SSY form). No copy was made either of the information on 'Changes of Occupation' or of 'Land Taken Over', which was added by St Anne's staff to the details on the Parish Lists but not required for the NFS record.

Form C51/SSY was sent to the relevant CWAEC, which in turn made a copy on blank forms supplied by St Anne's (or simply took from the C51/SSY any information required) and sent the 'original copy' to the appropriate Advisory Centre, being required to do so by December 1941.

Form 398/SS was now divided in two, and the part forming pages 1 and 2 was filed away at St Anne's, in accordance with Statistical Branch's regular system for dealing with the quarterly returns, to be kept for a period of 3 years in case of the need for further reference. The second part making up page 3, with the information on labour, motive power, rent, and length of occupation, details of which were not required by either St Anne's or the CWAECs, was sent directly to the appropriate Advisory Centre. The Statistical Branch at St Anne's started the process of sending the pages 2 and 3 forms to CWAECs and Advisory Economists on 12 September 1941. The work was completed in 2 months, being carried out in three rooms on the ground floor of the Sandy Knoll Hotel. There were 21 staff undertaking this work at first, including 18 women temporary clerks from the Poultry Section, but the figure later increased to 46. The women from the Poultry Section had been engaged previously on the farm survey work, and it is likely that it is their names and initials that appear on the C51/SSY page 2 copies.[53]

In this manner, the Advisory Economists received all the component parts of the Primary Record and the Census Return, but these records had still to be matched – the Primary Return for a particular farm had to be placed with the Census Return for that farm in order to complete the total record for each holding in the province of the Advisory Centre.

Completion of the maps

The completion of the map sheets was the responsibility of the CWAECs, not the Advisory Economists. The work was done to greatly differing standards. The maps produced by some counties are superb, while others are singularly scruffy and inadequate. The high standard of many of the maps shows that the work must have been undertaken by individuals with professional draughting skills. As few maps were signed (there were no instructions for this to be done) it is unlikely that these individuals will ever be identified, but without doubt their work represents one of the finest and most valuable legacies of the NFS. Certainly, the Land Commissioners showed an active interest in the work.[54] Later in the course of the survey, the CWAECs were instructed to make copies of the maps; they could retain the copies for their own purposes, but must send the originals to the Advisory Economists. Many CWAECs, however, appear to have retained the originals until they were told in May 1943 by the Ministry of Agriculture that the completed maps should be sent forthwith to the Advisory Centres, as their need for planning purposes was now urgent.

This was the part of the farm record that was the most useful to both CWAEC and Advisory Economist alike, and the high standard of many of the maps reflects the great interest shown in this aspect of the work, which often contrasts with the sometimes much poorer standard of the Primary Record. How exactly the CWAEC officers worked on these maps is not recorded, and we can only speculate on how the task was done. The work – first of obtaining all the necessary Ordnance Survey (OS) map sheets, and then, in the case of the 6-inch sheets, preparing them by transcribing all the OS parcel numbers to them from the corresponding 25-inch sheets that included this information – must have been most time-consuming and painstaking. Sometimes the transcription work is so meticulously done that it is hard to believe that all the tiny numbers are hand-written and not printed.

The representation of each farm, and the determining of its boundaries, must often have demanded repeated fieldwork visits. Some of the CWAECs appear to have maintained a record of each farm by a series of field slips (i.e. all the OS parcel numbers relevant to a particular holding), which would have enabled the various blocks of fields and rough grazing making up the farm to be determined on the map. Yet there must have been many details, in particular in regard to small or fragmented holdings that were difficult to show on a 6-inch map, or in the relation of holdings, for example, to adjoining common land or woodland, that could only be resolved by reference to the surveyors.

Then would come the job of showing each holding by a system of colour coding – at its finest by full colour washes applied with the expertise of an artist. Sometimes keys to the holdings would be given on the margins of the map, which demanded further delicate and painstaking work. Usually, however, the CWAEC cartographers transcribed the Ministry's farm identity code to the area of the holding on the face of the map. Sometimes, the full code was given, e.g. XE 303/59/46, but more often it was just the parish and farm number, 59/46. There are some counties where the Ministry code is not shown on the maps at all; instead they used their own coding system, bringing potential confusion when working with both the farm records and the maps. Thus, West Sussex used a system of case numbers on the maps, with a key to the farms and farmers on the margins. These case numbers, however, are also added to the Primary Returns. In instances where there was, in fact, no code reference to place on the map (usually because there had been no Census Return for that holding), then the name of the occupier, and sometimes the name of the farm, is added to the map face.

The maps were the last component of the farm record to be completed. For some counties this had still not been done as the war finished and it seems that in some areas the maps were never completed. By the end of 1941, only one or two completed maps had been sent to the Ministry of Agriculture for comment. A report on the progress of the survey issued in September 1942 stated that the mapping was 60% complete in 15 counties, 30–60% complete in 18 counties and under 30% complete in seven counties. In some seven counties as well, the maps required showing the farm boundaries were supplemented by plans of individual farms. The staff of one county (Surrey) misunderstood the instructions completely and initially supplied only maps of individual farms.[55]

By July 1943, about 66% of the fieldwork necessary for the maps had been completed. However, in that month, at a meeting of the Farm Survey Supervisory Committee, Dudley Stamp announced that some counties had hardly started on the mapping work.[56] The CWAECs were informed that the completion of the mapping was now urgent as the maps were in demand for planning purposes, and they should be transferred to the Advisory Centres as soon as possible. It was emphasized as well that the maps should not be taken into the field or used in a manner that might damage or deface them. Furthermore, they should not be brought up to date by the inclusion of later information, particularly relating to farm boundaries, as this would destroy comparability with the Primary Returns and the 4 June 1941 Census Returns. If the CWAECs wanted to show alterations in farm boundaries, they should use their own copies of the maps for the purpose.

In September 1943, the Ministry restated its concern about the maps becoming soiled, torn or otherwise defaced, noting that this was incompatible with one of the main objectives of the survey – the permanent Domesday record of the farms of the nation. At the same time the Ministry requested the CWAECs to accelerate the mapping, and if only about two-thirds was completed they were urged to appoint a part-time Maps Officer to speed the process. Priority was to be given to areas near large cities that had suffered from bombing damage.

The frequent use of the maps, and their borrowing by other bodies, had been anticipated as early as March 1942.[57] It was then hoped that the originals would not be damaged, and the Ministry was prepared to make the negatives available from the photographic reduction work so that further copies would be possible at the $12\frac{1}{2}$-inch scale. The split between the use by the CWAECs of 6-inch sheets and the 25-inch reduced sheets was about half and half. At the Conference of Advisory Economists held on 18 June 1942 it was said that the requirements by the CWAECs for the 25-inch reduced sheets might rise to 66% of the whole.[58] In fact 40,000 sheets at the 25-inch ($12\frac{1}{2}$-inch) scale were needed to cover the whole of England and Wales.

When the maps were transferred from the CWAECs to the Advisory Centres, they were stored in the special steel chests that the Ministry of

Agriculture had supplied for the purpose. Some of the Advisory Centres checked the quality of the maps they were receiving, in particular to see if the acreages of holdings matched the representation on the maps, and to determine whether the farms within a parish were complete and the OS parcel numbers accurate. However, the Advisory Centres were not formally required to do this work.

Interest in using the completed Farm Survey maps was considerable. The Ministry drew up rules in April 1943 about the disclosure of information from the National Farm Survey documents, which included rules on access to the maps. Officials of the Ministry's Land Planning Branch, and of other central government departments, had the right to scrutinize, copy and sometimes borrow the maps. 'Persons of semi-official status' might be allowed to examine the maps, but under no circumstances could they borrow them. There were also many post-war requests to use the maps, but the rules on access remained constant until 1956, when they were relaxed to allow students and research workers to study them 'under certain conditions'.[59]

Received at last by the Advisory Economists, the completed maps were now sent in instalments from the Advisory Centres to the Ministry of Agriculture and Fisheries' Planning Division (Cartographic Section) at 124 Cromwell Road, London, for transcription on to maps at the $2\frac{1}{2}$-inch scale. These were intended for use as master maps for land planning purposes, the main object being that there should be available in the Ministry of Agriculture a record of the boundaries of all farms and other holdings whenever proposals were made for housing and industrial development, and for conservation areas (green belts) around cities.

In January 1944, Stamp reported that the work of transferring the boundaries of farms on to $2\frac{1}{2}$-inch sheets would be supplemented by showing farm areas by colour washes. In that way, any unexplained areas, being white, 'would stand out at once'. Important information would be gained on common land, which would also appear as excluded from the boundaries of the farms. He showed examples completed for Wiltshire to members of the committee, telling them: 'These maps are of fundamental and outstanding importance for any work of rural planning and agricultural reconstruction'.[60]

The task of preparing the $2\frac{1}{2}$-inch map sheets was enormous: some 15,000 person days of work were estimated. Some counties (notably Norfolk), realizing the urgency and importance of the work, had undertaken it themselves. The first areas that were copied were those around bombed cities, where extensive rebuilding was likely to take place. The copying at the $2\frac{1}{2}$-inch scale in colour was continued by the Ministry's Land Use Division (based at Oxford Square, London) and by 1948 the maps for Cumberland, Lancashire, Soke of Peterborough, Norfolk, Suffolk, Essex, Hertfordshire, Middlesex, Surrey, East Sussex, Berkshire and Warwickshire had been completed, while those for West Sussex and Kent were in hand.

After their use for this work, the NFS maps were returned to the Advisory Centres.

In December 1942, J.H. Kirk proposed that CWAECs be given the task of updating the NFS maps in a programme of continuous revision.[61] It was felt that post-war rural land planning would go on for many years and up-to-date information would be constantly required. The revised maps should also be of great use to the CWAECs. But various objections regarding the impossibility of revising the different sets of maps, including Stamp's copies at the $2\frac{1}{2}$-inch scale, led to the proposal being dropped.

Notes

1   PRO MAF 38/208.
2   *Farmers Weekly*, Editor's Diary 12 July, 19 July, 26 July, 2 August, 4 October.
3   PRO MAF 38/208.
4   Copies of the forms used by Cheshire, together with a useful summary of the purpose of the first survey and the way it was carried out, are in PRO MAF 38/208 (FSC3). A copy of the form used by Leicestershire is in PRO MAF 38/213. Other examples of county forms used for the 1940 survey are included in PRO MAF Divisional Office records, e.g. Suffolk is at PRO MAF 157/1.
5   PRO MAF 38/208.
6   The county reports of the 1940 survey are in PRO MAF 38/213. Copies of the Ministry's summary report on these county reports 'prepared by Miss Skrimshire' are in PRO MAF 38/209 and PRO MAF 38/470.
7   Sir Donald Fergusson (1891–1963) was at the Treasury from 1919 and was promoted to Assistant Secretary in 1934. He was Permanent Secretary at the MAF from 1936 to 1945, retiring in 1952.
8   John Henry Kirk (b. 1907), whose influence on the NFS was considerable, joined the Ministry in 1934 and left in 1965 to take up the first Chair of Marketing at Wye College. He received the CBE in 1957. Still alive in 1995, he unfortunately died before he could be interviewed for this volume. An overview of agricultural change by him can be read in Kirk (1979). M.G. Kendall became a well-known statistician and historian of that discipline. See Kendall (1939), and also the preparatory and secret manuscript by Kendall prepared for the Ministry in PRO MAF 38/431. He states that his study was undertaken 'with a view to throwing some light on the areas which will best respond to stimulus in a campaign such as would be initiated on the outbreak of war'. See also Whitby (1946).
9   A list containing the addresses of some 30 of the CWAECs is in PRO MAF 38/575. Addresses can also be obtained from the original C47/SSY form in PRO MAF 32, and from PRO MAF 39/228–324.
10 Files on work of the Advisory Centres financed by the Development Fund are in PRO D 4.
11 The Development Fund, which had been established in 1909, was administered by the Development Commission, consisting of eight Commissioners appointed by Royal Warrant. It was at this time used to finance agricultural research and the agricultural advisory services. The first bid to the Development Fund for the NFS was for £14,000; of this £13,450 was agreed. A Development Commission file on the finances of the survey (D 4/226) survives as does a register listing the application of the MAF for a grant for the NFS (D 2/57).
12 Summaries from 1906 of the monthly crop reports for the whole of England and Wales are in PRO MAF 82: those for the Second World War period have the reference PRO MAF 82/196–197. From 1950 the reports are in full for each county, the Crop Reporter becoming known then as the County Advisory Officer. In 1959, the Monthly Crop Reports were termed Monthly Agricultural Reports.

13 PRO MAF 38/470 includes a letter of 30 December 1941 which states the importance of the Crop Reporter and the service he provided, saying it was very hard to recruit the right sort of professional man for the work.

14 Interview with the late Mr Nigel Harvey (15 February 1995) who worked with the Oxfordshire CWAEC during the war.

15 PRO MAF 38/210 (10); 38/207.

16 PRO MAF 38/209 (FSC10).

17 PRO MAF 38/205; these categories were largely repeated as the objectives of the Survey in the 1946 Summary Report *National Farm Survey of England and Wales (1941–1943)* (HMSO 1946), 2.

18 PRO MAF 38/209 (FSC13).

19 PRO MAF 38/215.

20 Dr Frank Yates, CBE, was Head of the Statistics Department at Rothamsted Experimental Station at Harpenden. He had joined Rothamsted in 1931 but became a scientific advisor to various ministries from 1939. With R.A. Fisher he had published *Design and Analysis of Factorial Experiments* (1937), and his *Statistical Tables for Biological, Medical and Agricultural Research* (1938) went to many subsequent editions.

21 The minutes of these meetings are in PRO MAF 38/207 and PRO MAF 38/210.

22 PRO MAF 38/209 (FSC5).

23 PRO MAF 38/207 (6).

24 An example of the confusion can be seen in the report by the Chairman of Kent WAEC, Lord Cornwallis, who directly compared the two series of gradings, and sought to explain the lower proportion of 'C' farmers in 1941/42 (8%, compared with 12% in 1940/41) by the fact that the CWAEC had taken over some of the holdings in the interim, and that education and 'fair prices' were playing their part. See the compendium of county reports in 'War-time food production: the work of War Agricultural Executive Committees', *Journal of the Royal Agricultural Society of England* 108 (1947) 80.

25 PRO MAF 38/209 (FSC9).

26 PRO MAF 38/209 (FSC13).

27 The press release is in PRO MAF 38/212 (14).

28 PRO MAF 38/207 (FSC15).

29 Blank sample copies of the census forms, stamped in red 'For Information', can be found in PRO MAF 38/209, PRO MAF 38/211, PRO MAF 38/214, PRO MAF 38/256, and PRO MAF 38/470. The forms in PRO MAF 38/256 are unstamped. The Parish Lists are preserved as PRO MAF 65.

30 PRO MAF 32 usually includes the originals of this form, but sometimes it is the copies made by the CWAECs that have been preserved.

31 This information does not form part of the NFS record in PRO MAF 32, but is included in the Parish Lists – PRO MAF 65.

32 PRO MAF 32 includes the detached p. 3 of the form. Very rarely, and noted only from Dorset records, the p. 1 form is also included in PRO MAF 32 (together with the C47/SSY) – this must be an error made at St Anne's, which evidently transferred the p. 1 sheets to the Southern Advisory Centre together with the p. 3 sheets. It is this apparently unique survival of the p.1 for Dorset parishes that enables comparative work with the C47/SSY to be made.

33 This form is included in PRO MAF 32.

34 This form is included in PRO MAF 32. There was also a smaller size of this form, dated 2/42, but otherwise complete and with the same reference.

35 An example of an M19524/5644 survives in PRO MAF 32/366/12 (schedule 31).

36 A possible example of such a form survives with the records of Cockfield in Co. Durham (3/2). This is on flimsier paper than usual, and has perforations at the top as if torn from a pad. There are also some slight differences in the questions under 'Conditions'.

37 Original and copied forms are in PRO MAF 32. An example from Wettenhall in Cheshire (183/10) is on grey paper.

38 PRO MAF 38/207 (1); 38/215; 38/210 (E1739); 38/208 (FSC5).

39 PRO MAF 38/208, PRO MAF 38/207 (7) and PRO MAF 38/210 (37c).

40 PRO MAF 38/210 (66–67).

41 PRO MAF 38/209 (4).

42 It is strange that the Primary Return form omitted a space for the acreage figure, and that this was never added at a subsequent reprinting – indeed, no later alterations were ever made to the form.

43 PRO MAF 38/207, 38/209, 38/210 (67), 38/210 (166a), 38/210 (200), and 38/211; for Form B501/E1 see PRO MAF 38/211 (56) and 38/209.

44 PRO MAF 38/210 (163).

45 PRO MAF 38/210 (E1739).

46 PRO MAF 38/209 (17).

47 PRO MAF 38/209.

48 PRO MAF 38/212 (21, EF393B).

49 PRO MAF 38/212 (14).

50 Work on the transcription of surveyors' information on to the Primary Record form at Sutton Bonnington on a 4-month temporary contract that began on 1 January 1941 was recalled in an interview with Miss Ann Nowill (28 July 1995).

51 PRO MAF 38/211 (73).

52 PRO MAF 38/211 (61).

53 The whole process is described in detail in PRO MAF 38/470 (SSY2612).

54 Some map sheets do seem to have been copied at the Advisory Centres – perhaps these were later copies. The South-Eastern Advisory Centre at Wye records that the copying of 6-inch sheets to sheets at the $12\frac{1}{2}$-inch scale was done by senior grammar school boys under the supervision of their masters. See PRO MAF 38/865(83B).

55 PRO MAF 38/207 (FSC14).

56 Dudley Stamp (1898–1966) was Reader in Economic Geography at the London School of Economics 1926–1945, and later Professor Emeritus. Director of the First Land Utilisation Survey of Britain, he was vice-chairman of the Scott Committee on Land Utilisation in Rural Areas 1941/42 and was Chief Adviser on rural land use to the MAF 1942–1955. Among many honours, he received the CBE in 1946 and was knighted in 1965.

57 PRO MAF 38/207 (43).

58 PRO MAF 38/212 (5).

59 PRO MAF 38/865 contains much correspondence relating to access to the maps, or requesting information from the maps.

60 PRO MAF 38/212 (RSC8) and 209 (RSC6).

61 PRO MAF 38/212 (EF393).

Contemporary Critiques:
Interpretation, Analysis and Policy   4

We now turn from the initiation and procedures for this huge state-sponsored data collection, and its progress, to examine more closely what contemporaries wished to do with the material. We then move on to an analysis of the problems associated with the assembled material, and provide a narrative account of the transferral of this material into the public domain.

Analysis and Publication

Papers distributed to the members of the Farm Survey Supervisory Committee ahead of its meeting on 5 March 1942 set out provisional plans for the analysis of the data collected for the National Farm Survey (NFS). There were to be two main types of analysis: a national analysis to be carried out by the Ministry, and a provincial analysis to be carried out by the Advisory Economists. It was proposed that the national analysis would be of a stratified 10% sample of the records and the provincial analysis would be of a similar 20% sample. In addition it was thought that various of the County War Agricultural Executive Committees (CWAECs) 'where the survey has been treated seriously and a good job done' should have the chance to carry out their own county analyses.[1] Indeed, these CWAECs had been pressing to be allowed to do this.

Analysis at the national scale

The national analysis was seen as falling into two parts: a summation of totals by county of 'the more important parts of the material', e.g. the pro-

© CAB International 2000. The National Farm Survey, 1941–1943
(B. Short, C. Watkins, W. Foot and P. Kinsman)

portion of farmers who were tenants, the number of full-time farmers, the number of farmers with piped water, etc.; and secondly an analysis that would necessitate the tabulation by punched cards of the various records making up the survey, using either the Hollerith or the Powers tabulating machine. However, because of the costs involved, it would be impossible to consider carding all the data made available by the survey. From June 1942, the Census Returns were due to be machine tabulated as a regular procedure, and consequently it was felt that there was no need to concentrate too closely on the 4 June 1941 material available with the NFS, other than on those questions on the page 3 Supplementary Form.

There were other points made on the proposed national analysis. There was interest in extending the NFS into a survey of landed estates, as noted above, which would show, in particular, the efficiency or otherwise of landowners in respect of their agricultural responsibilities. Owing to the large amount of work that would be required, this idea was given low priority 'unless the Ministry urgently requires it for administrative purposes'. There was also a proposal that holdings of fewer than 25 acres, and holdings where the occupier was neither a full-time nor part-time farmer, should be excluded from the analysis. In addition, it was considered that, as some 10% of the records of the total survey were in an incomplete or substantially inaccurate state, and as the work of upgrading these records would be extremely difficult and costly, they should be excluded from the analysis. Furthermore, some of the information that it might be desirable to tabulate was not in a state where it could be carded without considerable preliminary work. Thus the 4 June data did not include the agricultural area of the holding, the type of farm or the part of it that was arable – information that was needed to help to interpret the analysis. Finally there was a suggestion that the Advisory Economists could divide the country into types of farming areas and do the analysis by each such area, but this would be an immensely laborious task involving much preliminary work.

Among the proposals placed before the committee members was one, stated to be 'purely preliminary', for 'the number of tabulations and interrelations which may be worth exploring' and the design of a card for the Hollerith system. This proposal was drawn up by Kirk, Whitby and Yates acting as an informal sub-committee, and they expressed the hope that the NFS would do for the whole of England and Wales what Thomas and Elms (1938) had done in their Survey of Buckinghamshire Farms and Estates.[2]

The Farm Survey Supervisory Committee decided to set up a formal sub-committee (the Research Sub-Committee) to consider further the question of the analysis of the NFS data by machine carding, and 'to design the card and frame instructions for the transcription of material from the records and returns on to the card'. Dudley Stamp, a member of the Research Sub-Committee, asked for attention to be given to cartographical analysis of the data as well as to statistical analysis. The new Research Sub-Committee duly

Table 4.1. Proposed sample strategy for the National Farm Survey (NFS) analysis.

Acreage	Sample size (%)	Holding numbers
5–25	5	4,500
25–100	10	16,800
100–300	25	15,000
300–700	50	5,000
700+	100	1,000

met on 1 July 1942. Before the meeting a paper dated 24 June by Whitby, setting out further proposals for the method of analysis, was distributed to committee members. The paper noted that Hollerith machines were 'both possible and desirable'; that two 80-column cards should be used at the punching stage for reasons of cost and convenience, outweighing the fact that one of the two cards would eventually have to be wasted (Fig. 4.1). Cards cost 7s per 1000. Working on the basis of the availability of one machine only, the punching of all the records for the national analysis should take about 10 weeks, and the sorting and tabulating another 35–40 weeks. It was hoped that several machines would be available. The stratified sample was to be drawn up on the basis of holding size, as indicated in Table 4.1. The acreages of the selected holdings were to be measured by areas of crops and grass, excluding rough grazings.

At the meeting of the Research Sub-Committee on 1 July some Advisory Economists stated that they would like a straightforward random sample, rather than one calculated on a stratified basis. However, it was calculated that for a random sample to achieve comparable accuracy with a stratified sample, it would require three to four times the number of records, i.e. about 120,000 records instead of the proposed 42,000.

The need to test the data for the national analysis by running it through the Hollerith machines (the Treasury had approved the use of this system in May 1942) led to an approach to J. Wyllie, the Advisory Economist at the South-Eastern Agricultural College at Wye, that records of Kent, which was considered one of the counties most advanced in the survey, should be used for a pilot project. This would mean sampling some 6000 Kent records – the 4 June Census Returns as well as the Primary Returns. The stratified sample of these records would be selected in accordance with the plan set out by the Research Sub-Committee and it would represent about 15% of the total number of holdings above 5 acres in Kent (some 900 farms). The selection of the records would be facilitated by using an abacus, and 'We would see that one of these contraptions was made available.'

There were many problems in obtaining the necessary Kent records in good time for the pilot project, and alterations had to be made in the way the records were selected and the way that these were brought up to the required level of accuracy for the pilot analysis. In addition, it was decided to use another sample taken from records of the North Riding of Yorkshire. Yates of Rothamsted was in charge of the pilot project and his particular interest, skill and experience in the analytical processes was one of the major forces in seeing these stages of the survey through to a successful conclusion. In November 1944 he was recommended by Kirk for the award of an honorarium of 100 guineas (with 50 guineas to his colleague, Kempthorne). Yates's method of analysing the 15% sample was said to have saved £2000 of the anticipated costs. The award was not made.[3]

In September, 1942, the Ministry drew up its instructions for the selection of pilot samples for Kent and the North Riding of Yorkshire. The main

Fig. 4.1. A small Hollerith installation of the 1930s. Such machines were extensively employed in analysing the results of the National Farm Survey (NFS) (Source: Science and Society Picture Library National Museum of Science and Industry No. 1238/75 List B).

difference from the original proposals was that the 1000 or so selected records would be matched and brought up to the right standard after selection, and not from the pool of records available before matching, many of which might be in a poor state and perhaps never matched. The method of selection was either direct – i.e. by taking the 4 June returns, arranged by parish, and for one staff member to call out the acreages while another marked the selected records according to the percentages required by acreage size groups – or by using a copy of the Parish Lists (to be supplied by Statistical Branch or by Dr Yates). The former method was preferred, since the Parish Lists included holdings of fewer than 5 acres and did not refer exactly to 4 June 1941. The matching process and the problems that were likely to be encountered were very involved, and are described at length in the Ministry's correspondence.[4] Any records that could not be matched were to be included in the sample for the Ministry to make a final decision as to whether they should be retained or discarded. The last process was to annotate the record as a guide to the operators on the punching machines. Normally, this work would be done by the Advisory Economists' staff, but the Ministry was prepared to do it for the pilot sample.

The Farm Survey's progress report of September 1942, noted that the national analysis was to be based on a random sample of the records and not on all the 300,000 holdings that had been surveyed, but a final decision on the method would be reserved until after the findings of the 15% stratified random sample of Kent and the North Riding of Yorkshire could be analysed. Similarly, the Advisory Centres would be informed later regarding the method of annotating the matched records that were to be punched on the Hollerith cards.[5]

At a meeting of the Research Sub-Committee on 8 June 1943, the report on the trial sample for Kent, dated 28 May 1943, was considered. Kent records alone had been used for the mechanical tabulation process; those for the North Riding of Yorkshire had only been 90% complete and were used with those of Kent for a trial of the sampling procedure.[6]

Some committee members thought that the statistical material to be produced from the NFS data should be used, if possible, to help the wartime food production programme. The data should be summarized by county and, if possible, by CWAEC district. Yates thought it would be too difficult and costly to gain results by districts, although it might be possible to do something by merging districts into larger units. However, it was feared that the publication of county results might lead to ill-feeling amongst the CWAECs, with a possible subsequent adverse effect on the food programme. The intention was to prepare a national report as soon as possible that would deal with results for the whole country. The results for individual counties might have to be included, but as a statistical appendix only.

Attached to the full and comprehensive report on the pilot sample of records for Kent is a 'List of Counts and Tabulations' prepared by Hollerith for the analysis of the Kent pilot sample. There is also a list of suggestions for further analysis, which includes the possibility of the construction of maps of rent or rental value, of land fertility and of the classification of estates. The Kent pilot scheme did indeed serve to demonstrate that the Hollerith system of mechanical tabulation was suitable for the national analysis; that the use of two cards per farm was preferable to the use of a single card with double punching; that the sampling errors were not large; and that adequate results could be obtained by basing the analysis on a 15% sample.[7]

A paper prepared for the meeting of the Farm Survey Supervisory Committee on 26 July 1943 reported that the matching process to be carried out at the Advisory Centres had been preceded by the drawing of a 15% sample of the records for the national statistical analysis. It had been decided that, rather than wait for all the Census Return records to be matched with the Primary Returns (work that would take a great deal of time), the Advisory Economists should select their 15% beforehand (using the 4 June returns checked against the Parish Lists for this purpose) and then concentrate on the matching of this 15% sample. 'While ideally perhaps the matching of record and return should have preceded the selection of the sample, had this method been adopted the national analysis would have been further delayed by at least six months.' The rest of the matching would be left 'until a later date'.[8]

The coding scheme to allow the data from the Primary Returns and Census Returns that were selected for analysis to be punched on the Hollerith cards was altered from that which had been used for the Kent pilot sample, which was considered to have been 'on the intricate side'. It also needed a particular Hollerith machine to operate it and this might not be available later. Consequently, Yates and his colleague, Kempthorne, devised a second coding scheme entitled, 'Final Instructions for Coding', which was issued to all Advisory Economists in April 1943.[9]

The data to be punched on the cards was transmitted to the Hollerith machine operators by the use of coding sheets, which were double-sided, with information from the Primary Return on one side and from the Census Return on the other. Some 40,000 coding sheets were to be forwarded to the operators in batches of some 8000–10,000 sheets. The first batch was ready at the end of July 1943, and the others followed at monthly intervals. By 15 July 1943, completed coding sheets had been received for Kent, Surrey, Wiltshire, Northumberland, West Sussex, Cheshire and Montgomery, and those for Oxfordshire, Staffordshire, the East and North Ridings of Yorkshire, Lincolnshire (Kesteven) and Devon were expected shortly. At the same time that the Advisory Economists submitted their coding sheets, they included a list of rejected records and returns, with reasons for rejection. The estimate was that the number of rejected records would amount to 1–2% of the total sample of 40,000.

In order that the Advisory Economists understood the coding instructions properly, the Ministry of Agriculture invited the Farm Survey supervisor from each Advisory Centre to come to London for a conference on two occasions. These conferences enabled the discussion of problems and the clearing up of misconceptions. The Ministry kept a check on how well the coding was being done by calling for 20 original records from the sample which could be scrutinized by its staff. The coding itself seems to have presented little difficulty: the main problem was in maintaining complete accuracy in what was a slow, monotonous job.

Owing to pressure by all national users on the Hollerith tabulating capacity, the Research Sub-Committee had limited their programme of statistical analysis to 'those counts and tabulations of which there is a good assurance that the results will be interesting and useable'. The NFS had the Hollerith machines available for 6 months to accomplish the necessary work. The figures that emerged from the process of mechanical tabulation were in a crude state, and these needed to be refined into readable tables. This work would take a year and involve the full-time services of two qualified statisticians, five clerical workers (including a supervisor) and four electrical calculating machines. Rothamsted Experimental Station offered to do the major part of the work under Yates's general supervision. It was expected that results would begin to emerge at intervals, starting in September or October 1943.

Regional analysis

The Advisory Economists would also be undertaking research projects into the NFS material. In August 1942, the Ministry had told the Advisory Economists that they should consider planning a programme of provincial analysis of the records in their custody. Kirk considered this provincial analysis would be of great benefit in supplementing the national analysis and that it would be difficult to restrict the Advisory Centres as to the type of analysis they carried out.

> I can ... visualise the University of Cambridge protesting vigorously if, after their Advisory Economist has done a great deal of low grade work on the Survey, he were told that he could not use it for research but must regard himself as a mere collector of material for others to research upon.[10]

One of his concerns was that the Advisory Economists should not publish the results of their work ahead of the Ministry's report on the national analysis, and that all provincial work should be passed to the Ministry for approval. A letter to Advisory Economists making this point was therefore issued on 29 August 1942.

Early in 1942, the University of Bristol had carried out various experiments into the NFS data they were receiving. The University was thanked by Whitby of the Ministry of Agriculture for the results: 'This sort of preliminary analysis is of considerable interest to us, both from the point of view of the kind of analysis, but more particularly from the view point of providing a test of internal consistency for the material.'[11]

On 2 November 1943, Kirk asked the Advisory Economists to present their plans for the analytical work at the next Conference of Advisory Economists, which was held on 14 December 1943.[12] These research programmes were summarized in April 1945 in an application to the Treasury for a further grant of £10,000 from the Development Fund for the work to be completed (Box 4.1).

The proposals from Aberystwyth and Reading were considered too ambitious by the Ministry and were cut back. The lack of a research proposal from Leeds was at least partly due to the fact that one of W.H. Long's staff, Dawson (possibly the farm survey supervisor), objected to doing any research that was not directly connected with the war effort. W.H. Long added to his letter giving this information to J.H. Kirk: 'I fancy Dawson's complaint is symptomatic of the feeling amongst the Survey workers in Provinces other than this.' Kirk replied: 'You do seem to have your full quota of men with active consciences, and although one cannot but admire their attitude, it is fortunate from an administrative point of view that the problem is more or less confined to your province.' In fact Long, with the help of G.M. Davies, wrote his economic survey of Swaledale farming in 1948, using the NFS material and including material on farming based on the 1941 4 June Returns (Long and Davies, 1948).

A memorandum by R.R. Enfield of 10 February 1944, states:

> In the majority of cases the proposals for investigation of the Survey material at the provincial centres are of the kind which would produce a good deal of interesting information, interesting particularly to the advisory centres themselves, to students and in some cases to persons concerned with practical questions of agriculture, estate management etc., but I am bound to say I think investigations of the kind here described would be only of limited interest to this Department. In any case, I understand the Treasury have agreed to an expenditure of £25,000 to cover the whole of the provincial investigations into the Farm Survey material and I do not think we could reasonably ask for more. We should have to allot the grant with reasonable fairness between the various provinces, and, looked at from this point of view, it seems to me that one or two of the provincial proposals are too ambitious, notably Reading and Aberystwyth. It seems to me that we shall have to tell these centres that there is a limit to the money available and they will have to adjust their programmes accordingly.[13]

Box 4.1. Projected analyses of the National Farm Survey (NFS) data by the Advisory Economists, 1943.

Aberystwyth
- An analysis of the following factors in all the records in Wales, by parish, district and county: (i) ownership and tenure; (ii) type of farmer; (iii) grade of management; (iv) size of farm; and (v) rental values.
- A comparison of the results obtained with all the results shown by the county samples taken for the purpose of the national analysis.

Bristol
- About two-thirds of Herefordshire would be intensively surveyed, work in parallel with the proposed Ley Farming Investigation and Farm Management Survey. About 1500 farms would be considered. Material would be assembled from 4 June returns, from farm survey schedules and from farm survey maps.
- A study of the economics of central Somerset. It was proposed to widen the material of the NFS by incorporating statistics of milk yields and to study the relationship between farm size, stock carried and milk produced under various soil conditions.

Cambridge
- A survey of land ownership with the object of ascertaining: (i) the distribution of estates by size in the eastern counties; (ii) the area of land in owner's occupation; (iii) the effect, if any, of multiple ownership on the layout of the farm and condition of permanent equipment; (iv) the influence of owner-occupation versus tenancy on permanent equipment and condition of farm; and (v) the distribution of ownership by soil type.
- An investigation on the 'type of farm' – the distribution of farm size, and the utilization and condition of land. This would entail the analysis of the Primary and Census Returns for a sample of 5500 farms.

Harper-Adams
- An intensive study of the condition of farms and farming in about twenty parishes of south-west Shropshire. There would be an attempt to prepare a productivity map of the area using both 4 June data and material from the Farm Management Survey.
- A study designed to show the effect of the scale of land ownership on the condition of farms, looking at estates of over 5000 acres, estates of less than 5000 acres and at farms in owner-occupation. The analysis would involve all the factors under sections B, C and D of the Primary survey record, as well as two sections concerning Rent and Length of Occupation of the page 3 Census Return.
- An analysis of the survey material for Shropshire and Staffordshire by local type of farming districts.

Box 4.1. (cont.)

Newcastle
- A study of three areas, comprising 500 farms in all, for which economic and financial data are available from the Farm Management Survey to supplement the farm survey material. The three areas were the Orton–Tobay–Ravenstonedale area in Westmorland; the South Northumberland feeding area; and the Belford area in north-east Northumberland.

Leeds
- No research work proposed.

Manchester
- An analysis of land tenures, with particular reference to large estates in Cheshire and Lancashire.

Midland
- A summary of land occupation and ownership in Nottinghamshire. This would deal with the size and numbers of different land parcels comprising farm units and should provide interesting new data on the land tenure picture of agriculture in the area.

Reading
- A series of county reports (much on the lines of 'The Farms and Estates of Buckinghamshire', 1938) including:
  - (a) The compilation of terriers of all the estates in each county of the Southern Province, which will form the basis of a comprehensive study of ownership and estate-management.
  - (b) A recasting of 'size of farm' analysis on the basis of the size of the financial unit, to be followed by a study of the ways in which financial units have been constituted.
  - (c) A study of the present layout of farms, supplemented by a limited study of the historical development of such layout.
  - (d) A study of the economic types making up the farm population to comprise: (i) an analysis of landlords; (ii) an analysis of farmers; and (iii) an attempt to assess the numerical significance of the purely family farm.

Seale-Hayne
- An analysis of land ownership and land tenure related to the permanent equipment of the land.
- Classification of types and sizes of farms within certain defined areas.
- A further study of farm layout where this is described as unsatisfactory.

Wye
- Investigational work on: (i) farm layouts, and (ii) multiple holdings.

In fact, a further £10,000 was requested in April 1945. J.H. Kirk had written on 22 October 1943:

> There has, I believe, been from the outset of the farm survey, an understanding with the Centres that, having undertaken the labour during two or three years of assembling, checking and filing the Survey material, the Advisory Economists should have the use of it for research. It is to my mind doubtful whether the Economists or the Universities would have consented to undertake this drudgery except on those terms.[14]

Cartographic analysis

At a meeting of the Research Sub-Committee on 12 July 1943, Dudley Stamp put forward proposals for a trial cartographic analysis of a selection of items from the Primary Record, with the idea that this analysis should complement the pilot statistical analysis that had been done for Kent. Particular data from the Primary Records, such as the availability of electricity, water supply, the condition of buildings, the classes of farmers, the types of soil and the natural fertility of the farm, would be transcribed on to maps by a series of 'circle charts'. This method of approach was expected to be of great use in, for example, the planning of electricity and water supplies to farms, and would also have a considerable general research value. There was to be an experimental area for the cartographic project which had yet to be chosen. However, Stamp had suggested East Kent, and Mr Wyllie, the Advisory Economist for the South Eastern Province, had been asked to send 2000 NFS records to Stamp at the offices of the Ministry of Agriculture's Planning Division in Cromwell Road, London. The Executive Officer for the Kent CWAEC had also been instructed to give priority to the completion of the farm survey maps of East Kent. There was considerable resistance by both Wyllie and his farm survey supervisor, A.C. Baker, to letting these records out of their hands. They had already had the trouble of preparing the sample for the pilot statistical analysis, and they stated that they now needed all the records to bring them to a state of completeness, and also for other possible administrative purposes. They felt that any work on transferring data to maps could be carried out at their offices as they did not want to lose the maps for a period of up to a year. It was also pointed out that a large part of East Kent had been occupied by the military and that agricultural conditions there were consequently abnormal.[15]

Stamp wrote again on the cartographical analysis of the farm survey in a report to the Research Sub-Committee dated 21 December 1943. It had proved impossible to use the records of Kent for this project 'as the maps were far from complete'. The objections of the Wye Advisory Centre must have also been of considerable weight in this decision. Instead, the Swindon area of Wiltshire (District no. 1) had been chosen, as these farm records were believed to be substantially complete, and in due course Stamp received the relevant Primary Records.[16]

It had been decided to represent cartographically all the recorded details from the Primary Return, and then by a study of the maps to decide which could be seen to be of general or of permanent value. Each farm was identified, and the position of the farmstead marked with a dot on a 1-inch Ordnance Survey sheet, which was to form the working base map. However, as Stamp wrote: 'It was perturbing to find that even in an area believed to be complete about 100 farms were missing from the Primary Records.' The maps were next further prepared to show geology, relief, drainage, parish boundaries, roads and railways. Then the farm data was mapped, 44 separate maps being annotated with appropriate symbols for the site occupied by the farmhouse and for the answers to the various categories of questions, each map representing one particular question. The 44 maps were arranged in the order of the questions on the Primary Return. Stamp's report commented on the usefulness of each map and drew attention to particular findings, such as the degree of correlation between the condition of a farm and its geology, its situation with reference to railway and road and the fact that there was evidence that the answers on the Primary Returns were not accurate in this respect, that 'bad' farmhouses were on clay in the sample, that infestation with rabbits and moles coincided with areas of woodland, and that the different categories of farmers were found indiscriminately on all types of land.

A fourth meeting of the Research Sub-Committee was held on 27 January 1944. It considered Stamp's report, and decided that a list of priorities for the cartographic representation of the NFS should be drawn up. Top priority was to be given to the map illustrating the 'Class of Farmer', followed by that showing farms with or without an electricity supply. Five priorities, which consisted of 22 groups of maps, were eventually decided upon. It was estimated that to undertake the work needed to produce 12 of these maps for the whole country, a staff of some '4–6 Juniors and 1 Senior' would be required. The other work of transferring farm boundaries from 6-inch and $12\frac{1}{2}$-inch sheets to $2\frac{1}{2}$-inch sheets could be done by the existing staff of the Ministry's Planning Division.

Some of the CWAECs were said to be interested in the cartographic representation of tenure to show the distribution of estates, and indeed such maps for Lancashire had already been prepared. In addition, there had been some consideration given to undertaking a map that would depict an analysis of farm rent, and a sample of such a map was to be prepared by Yates' staff at Rothamsted.

Publication

In July 1943, thought was already being given by the Ministry of Agriculture to the final form of the presentation of the NFS data. A national summary statistical report was anticipated, with the main emphasis on factual information, and only a few interrelationships would appear. The national report would have county appendices which could be handed to the CWAECs, if desired.[17] It was also anticipated that the national summary report would be succeeded by a series of studies of particular subjects, such as tenure and occupancy, management and equipment. The preparation of these subject reports would mean the appointment of additional technical staff by the Ministry, and arrangements for this were already in hand.

Once the statistical analysis of the stratified sample of the farms of England and Wales was completed, the results were written up in a report prepared for publication. A draft of this report, known as the *Summary Report of the Statistical Results of the War-Time Farm Survey of England and Wales*, was submitted to the members of the Farm Survey Supervisory Committee and to the Advisory Economists for comment. The report had been prepared by the Economics Branch of the Ministry of Agriculture and Fisheries (MAF). It may have been written by Whitby: he was certainly the Ministry officer soliciting the views of the Economists and the members of the Farm Survey Supervisory Committee.[18] The draft includes two paragraphs and a table concerned with the thorny question of managerial efficiency and grading that were later cut from the final publication.[19] R.R. Enfield had written of the inadvisability of allowing these paragraphs to remain in the final text:

> partly for the reason that it might cause offence, partly for the reason that the grading of farmers was done primarily for wartime purposes and was of transient interest, and partly because if the Press or the National Farmers' Union happen to fasten on this, it might detract from the great importance of the useful and illuminating information in the rest of the Report.[20]

The report, now entitled *National Farm Survey of England and Wales (1941–1943): a Summary Report*, was eventually published by HMSO on 19 August 1946, with a Ministry of Agriculture press release giving the background to the survey and emphasizing its principal findings[21]. The Minister, now Tom Williams, wrote of the report, 'I agree that this is an excellent piece of work which reflects great credit on all those who worked to such good purpose to provide and analyse this valuable information.'[22] By 31 December 1946, 8125 copies had been printed and 6200 copies sold.

In September 1944, the Water Supplies Branch of the MAF had issued a memorandum to the CWAECs which stated that the NFS had shown that some 180,000 farms were without a piped water supply to the farmhouse and 210,000 without a piped water supply to the farm buildings.[23] Data from the survey was extended to produce a *Report on Farm Water and Electricity Supplies* (No. 23 in the series of Reports on the Economic Position of Agriculture) issued in September 1945.[24] The two reports on water and electricity had, in fact, originally been written separately, the latter by Miss Wilson – probably one of Whitby's staff. The introduction to the combined report states that it is 'the first of three which are based on material obtained from the Farm Survey of England and Wales carried out during the period 1941–43'.

Problems with the Survey

Problems with the Primary Returns and Census Returns

A memorandum from A.C. Baker, the farm survey supervisor for the South-Eastern Province based at Wye, set out the basic causes of the problems that beset the NFS virtually from its inception in April 1941. He wrote:

> The two fundamental difficulties in the scheme are (a) the endeavour to combine in one statistical record exact quantitative information with indefinite qualitative information which is a matter of degree and of opinion; and (b) the endeavour to match the basic record of the revisionary survey [the Primary Record], which is collected by personal inspection, and very exactly revised by the County Officers and by this office, with three separate returns, made under the supervision of a different authority; a return incompletely rendered and incompletely revised.[25]

From these fundamental flaws in the planning of the NFS arose many difficulties in the gathering, checking and matching of the records and their individual items of data that set the completion of the survey back many months, cost a great amount of extra money in the attempts to rectify the problems, and nearly led to the curtailment of the whole survey with its original objectives far from completed. In October 1942, W.R. Black and J.H. Kirk, among the most senior Ministry of Agriculture officials responsible for administering the survey, discussed the possibility of curtailing the survey and limiting the remaining work to a 15% sample of the agricultural hold-ings.[26] They were particularly alarmed by the spiralling costs, the unsatis-factory nature of some of the data and the slow progress towards completion. It was decided, however, to continue with the full survey and to try to resolve the various problems in the quickest and most convenient manner, but without compromising the validity of the data that was at the centre of the various disputes.

By September 1941, it had become clear that the proportion of 4 June 1941 Census Returns which initially could be judged as reasonably complete and accurate was not much above 50%. For NFS purposes this percentage was not acceptable, and further work on some 150,000 returns was therefore necessary. It was estimated that there were perhaps half a million errors to correct!

Most errors were on page 3 of the 398/SS form, which contained the additional questions for NFS purposes – questions that were unfamiliar to the farmers. Many of these supplementary forms were returned blank, or were completed in a manner indicating that the questions had not been fully understood. It was estimated that 60% were defective in some way. Occasionally, farmers refused to complete the questions at all, saying that they were not needed for the business of raising food production and that the Ministry had no right to be asking them. Some 60,000 reminder forms (C69/SSY) had to be sent out by the Statistical Branch. These reminders in turn generated further problems, for they resulted in several thousand written queries from farmers.

On 19 July 1941, A.B. Bartlett, head of the Statistical Branch, sent a worried letter to R.R. Enfield at the Ministry's head office in London about the pressure of work caused by the NFS and, in particular, about the errors found on the page 3 Supplementary Form. As the Statistical Branch was not in a position to sort out the problems arising from these errors, and because they were under such pressure with the September quarterly return by now overlapping with that of June, it was decided in September 1941 to hand over the page 3 queries concerned with the questions on 'Rent' and 'Length of Occupation' to the Advisory Economists. It was estimated that each Advisory Centre would receive initially some 300–500 of the written queries from the farmers, with the bulk of the incomplete page 3s following later. No extra finance for additional clerical help was offered to the Advisory Economists for this work to be undertaken.

By the time they received these page 3 queries, the Advisory Centres were inundated with the general work of the survey and with other particular problems. There was little headway that they could make with resolving the queries, and the NFS was consequently endangered by the sheer scale of the problems and the difficulties in resolving them. Indeed, Kirk suggested in a circular letter of December 1941 to the Advisory Economists that a percentage of holdings in each province, incorporating the poorest records, might have to be excluded from the final analysis.

As early as October 1941, the possibility had been raised of using the Crop Reporters to answer the page 3 queries. At first there was resistance to this idea, mainly on the grounds of expense, but by December the idea was being actively considered by the Ministry of Agriculture. In January 1942, the Treasury was approached to sanction the expenditure of a further £10,000 to cover payments to Crop Reporters. A letter of 22 January 1942 from R.R. Enfield to H. Biggs at the Treasury states that without the assistance of the Crop Reporters, and the necessary expenditure, a 40% deficiency in the page 3 answers would 'seriously risk the failure of the survey as a whole'. Biggs responded in a taut note, commenting upon the general expenses of the survey which had not been correctly presented to the Treasury:

> I am afraid that if I had had any idea of this I would not myself have approved the project, but would have submitted it for decision higher up in the Treasury. ... Now I understand a further £10,000 is required because the farmers have failed to fill in the 1941 June 4th Return (with its additional complications) properly.

Reluctantly, however, he agreed to the additional expenditure.[27]

It is interesting to reflect that, if only the questions that were consigned to the Supplementary Form (the page 3 of the 398/SS) could have been kept as part of the Primary Return and been answered by direct questions to the farmers during the course of the survey work by the CWAECs, then it is likely that only a fraction of the resulting muddle and expense would have occurred.

A memorandum was issued on 16 February 1942 from the Statistical Branch to Crop Reporters, informing the latter of the problem with the Supplementary Form questions. It was hoped that the majority of queries could be cleared up from the Crop Reporters' local knowledge, or by telephone or letter (in such cases the Crop Reporter would be paid 1s for each completed schedule). If the queries could be resolved only by a personal visit, then the payment would be 2s. At a later date, it was decided to add a travelling payment of a further 1s per farm to the 2s fee for farm visits. The completion date for clearing up the queries was given as 15 April, 1942. Some Crop Reporters complained that the scale of payments was not high enough, and a few refused to undertake this work. In such cases, there was no alternative to the Advisory Centre having to attempt to clear up the queries unaided.

The Crop Reporter was advised that farmers should be told that they were not legally obliged to answer the supplementary questions, but that their assistance would be of the greatest value to the MAF. Information supplied was to be treated as confidential and would not be disclosed to private individuals or to any organization other than the Ministry, the Advisory Economists and the CWAECs. If a farmer refused to answer, then the Crop Reporter was told not to pursue the matter 'to the bitter end' but to try to make an estimate of the answer from his local knowledge. This advice stemmed from the fact that the Ministry had originally considered referring some cases of non-completion of the 4 June return for prosecution, and a list of these cases had been drawn up in September 1941, but it had been advised by its Legal Department that it was unlikely that a prosecution would succeed.[28]

The Crop Reporters were sent by the Advisory Centres the whole dossier of papers relating to the page 3 queries, including the defective

white (or buff) form, C69/SSY, or in some cases the yellow/green version
on thicker paper of the 398/SS form – these were marked with red crosses
against the questions that required an answer. At least one Crop Reporter
(in Yorkshire) attempted to resolve the 600 defective schedules he had been
sent in an original manner. He estimated that visits to 600 farms would take
him 3 months. He speeded up the process by obtaining at least some of the
required information by visiting local garages, instrument makers and black-
smiths (for the questions on Motive Power); and land agents, valuers, etc.
(for the questions on Rent).[29]

At first it had been proposed that Crop Reporters should clear up all
queries, including those where the form was simply left blank. Later it was
realized that a blank answer often meant a correct answer (there were no
instructions to the farmer to write in 'nil', 'none' or 'not applicable'), and that
in cases such as these the Crop Reporter would be pocketing his fee for a
non-existent query. The Crop Reporter was asked to use his judgement and
common sense on such matters, but also to work with the guideline that, if
a farm was less than 50 acres, a blank return for Motive Power could be
assumed to represent 'none'.

By 30 April 1942, out of the 54,500 forms with queries that had been
received by Crop Reporters, 32,300 had been cleared up without the need
for travelling and 5200 resolved by visits to the farms. By 31 March 1943,
92,199 such forms had been sent to Crop Reporters and 74,158 had been
completed, leaving 21,594 forms with queries still to be settled at this date.
Some 60,000 payments to Crop Reporters were made at the 1s rate, and
10,000 farm visits had triggered the 2s rate (with additional costs of £514 for
travel). At 31 March 1942, the total expense of using Crop Reporters to clear
up the page 3 queries was £4547 16s 3d, and even with the final figure ris-
ing to c. £7000 this was well within the budget of £10,000.

Problems with the completion of the 4 June 1941 Census Return did not
just relate to the page 3 form. The C47/SSY form returned by the farmers to
the CWAECs was found, when checks were made, to be often 'far from
identical' with the page 1 of the 398/SS form returned to Statistics Branch,
in particular with respect to the figures for livestock, rough grazings and
permanent pasture. It became clear that some farmers were deliberately
returning higher figures of dairy cows to CWAECs in order to qualify for
greater allotments of animal feedstuffs, as well as overstating their acreage
of rough grazing as opposed to pasture in order to discourage further
plough-up.[30]

Problems also affected the Primary Return. The Advisory Centre
Recorders found that when copying the Primary Record in many cases it
was substantially incomplete. This resulted in many records being returned
to the CWAECs, necessitating re-surveys where necessary. In some areas, as
many as 90% of the Primary Returns were defective.[31] The problems were
various. There was a tendency to overgrade farmers, since an 'A' or 'B'

managerial rating meant that subsequent visits to the farm were less likely to be demanded. Also, with an 'A' grading there was no need to report on personal failings, which some surveyors were reluctant to do – at least partly because of the worry that they might be sued for libel if farmers found out what had been written about them.[32] Landowners, in particular, as they learnt about the survey, wished to have a greater role in the supervision of their tenants in the interest of increased food production, and sought to know the grading of their farms. The Ministry issued a memorandum on 5 February 1942 to Land Commissioners which made it clear that the farm classification should not be divulged to anyone other than the farmer himself, unless there was some action demanded by the landlord in the interests of food production, and even then the actual revealing of the grading category should be avoided if possible. On the question of libel, the Ministry said that, in cases where the grading was passed on to another responsible party, the CWAECs could claim 'qualified privilege', i.e. that the passing of the information was legally excusable in the circumstances. However, it added that the main defence against a charge of libel was to ensure the confidentiality of the information. There is no evidence that any such libel claim was ever laid against the CWAECs, or even contemplated.

In some cases incomplete returns might result from a calculated or principled refusal to answer particular questions by some farmers. The prolific agricultural writer A.G. Street, for example, refused to answer Question 3, asking whether he was a full-time, part-time, hobby or other type of farmer and whether he had any other occupation[33]. In his book *Hitler's Whistle* (1943: 271), he wrote:

> while I am meek enough and sensible enough to obey orders from experts concerning things of which I have no experience, the national farm survey form that arrived here the other day did not go down very well. I am more than willing to accept and to try to help the wartime control of my farming, but I'm damned if I will ever willingly submit to any interference with my personal life by this means. So across the questions … I wrote that I refused to answer as a matter of principle, and signed my name; much to the amusement and, I think, the joy, of the official who brought the form and inspected the farm. … I had always thought that we were fighting this war to preserve some freedom of thought and belief to the individual … not until it can be proved that his land is badly farmed and neglected should the question of his other activities and interests come into the case. … I hope one or two of my farming friends will visit me in gaol on visiting days.

A further problem with the Primary Return lay in the fact that some of the surveyors were the agents of the landowners, and this could mean that their standards of judgement were not always sufficiently detached. There was also a general lack of interest or energy on the part of the surveyors, which resulted in numerous errors. It was mainly these errors and omissions that resulted in the high return rate of the Primary Returns to the CWAECs.

But the principal reason for the lack of enthusiasm for the survey work by some CWAECs was the fact that they saw this work as largely unnecessary, and not directly connected with the real work of increasing food production. It was not that they did not think surveys were important – they had their own systems in place for monitoring the farms for which they were responsible, and indeed, the 1940 survey had helped set up such systems. The NFS, however, seemed an exercise in bureaucracy, collating data to fit various forms for purposes that did not seem directly connected with the war effort but rather more concerned with post-war planning.

Variations in the completion of the Primary Return by the counties within an Advisory Province can be seen from a summary of the situation in the South Eastern area in November, 1941. Baker wrote of the four counties of which he was the supervisor:

> *Kent*: The work of the survey is done mainly by the area committees with some assistance from the technical officers. The general comments often give a very complete and vivid picture of the farm. The farm surveyors vary from Lord Northbourne and Mr Garrad, the County Agricultural Officer, to some shrewd working farmers.[34] The result will be, I believe, of value for the present and of special interest for the historian of social conditions of the future.
> *Surrey*: The work is almost entirely done by the county technical staff and is more uniform and formal, and less vivid, than in Kent.
> *West Sussex*: General comments are very brief or absent altogether. I have mentioned to the officer in charge that W. Sussex differs from the other counties in this respect, but that is the business of the Executive Committee and I cannot go further than to point out the difference.
> *East Sussex*: The survey is being done mainly by specially recruited officers with technical agricultural qualifications, and is being used by the Executive Committee for their own purposes, as no complete survey was made before. General comments often go into greater detail than would normally be required and contain such notes as 'rabbits on the boundary have been reported to the Pests Officer'.[35]

Further problems with the Ministry of Agriculture's consideration of the Primary Return lay not with poorly completed forms, but with the opposite: that is, with survey work carried out so methodically that it was well above the Ministry's stated standards. What was to be the Ministry's response to CWAEC surveyors who were eager to know the use to which their hard work, over and above the call of duty, would be put? A good example is provided by the case of F.W. Bateson, the Survey Officer for the Buckinghamshire WAEC, who wrote to Enfield at the Ministry on 12 January 1942:

> What is really worrying me is the possibility that much of our carefully collected Bucks material may never be used. I expect you know that in some respects our original survey card was much fuller than your form. One of the

points I have insisted on throughout is that every single field on each farm must be inspected, described and graded. And this means that we can tell you *inter alia* exactly how many acres there are in the county of 'good', 'useful', 'moderate' and 'poor' grass, and exactly where they are to be found. ... Another item on which we shall soon have precise and complete information is the question of farm lay-outs. As we proceed with the marking in of the farm boundaries, we are listing the sizes and shapes of the fields, the position of the farmstead, the general shape of the holding and the number of detached blocks of fields. We can also supplement your form on such matters as the rotation of crops, the ages and occupations of our farm-workers, the acreages and lessors of grasskeeping land, the location of rabbit warrens, and of course milk-yields. As far as Bucks is concerned, I feel that the Survey provides an opportunity and an excuse to do a really first-class job. Will your 'local interpretation' of the Survey's results give me my chance?[36]

This was the very antithesis of the low morale and lack of enthusiasm for the NFS described above. Bateson was invited to the Ministry:

It would ... be of great benefit to us to have the advantage of any suggestions that you have to make based on your experience in carrying out the Survey. ... It is far from our intentions to boil the whole thing down to a kind of statistical summary.[37]

However, that is exactly what did happen, and there is no evidence that the detailed surveys for which Bateson was responsible were used for any purpose of the Ministry of Agriculture other than for the immediate wartime work of the Buckinghamshire CWAEC.

A further problem for the NFS lay with the general wartime conditions that resulted in the requisitioning of agricultural land by the military and which was constantly altering the acreages of holdings or even obliterating farms entirely. It was decided that military requisitioned land, where an entire farm (or a large part of it) was involved, would not be included in the national analysis.

The process of matching the Primary Return for each farm with its appropriate Census Return was held up until many of the problems set out above had been resolved. When the matching did commence at the Advisory Centres, it threw up its own range of problems. It might be instructive at this stage to consider the overall types of concerns being experienced at this time by the Advisory Centres as they sought to keep the vast amount of data they were receiving under the right type of control. Box 4.2 therefore contains a typical assembly of queries, anomalies, omissions and general concerns that might be concurrent at any one of the Advisory Centres.[38]

The main problems from the matching process arose out of the fact that the total farm record was built up of two different sets of forms that had originated with two different organizations – the Ministry's Statistical Branch and the CWAECs. In addition, the Advisory Centres formed a third body with different views and interpretations.

Box 4.2. Problems with the National Farm Survey (NFS) forms.

- Parcels of C47/SSY forms arriving from the CWAECs 'in no particular order'.
- Primary Returns that were incomplete and with obvious inconsistencies in the management gradings – some parishes all 'A' farms.
- Page 3 question 'Length of Occupancy' answered: 'I was born here'; 'All my life'; 'Since the 15th century'.
- Eleven C69/SSY forms received, all of which had been sent to one farm manager with no reference as to which referred to what part of his amalgamated holding.
- Forms with a schedule, but no address other than that of the Crop Reporter; forms with no addresses at all; forms with an address but no schedule.
- The page 3 'Motive Power' question on number of field tractors answered with the tractor engine number – 'an error which might give startling statistical results'.
- Three large parcels received but without any covering letter.
- Reaction received from farmers to reminder forms: 'This is not a small holding but a private residence, so your questions do not apply. Haven't you got anything better to do than send out these forms?'
- A parcel of C69/SSY forms received from Statistical Branch, poorly packed and with no names and addresses on the forms.
- Census returns which cannot be matched with Primary Returns: page 1s with no page 2s or page 3s; page 3s with no Primary Return …
- Comment at the Advisory Centre that the work is 500% more onerous than anticipated [!] (Seale-Hayne).

The principal difficulties were fourfold. Firstly, there was the problem of the interpretation by the CWAECs of what constituted a holding, and how this matched with the holdings shown in the Ministry's Parish Lists used for the Census Returns. Secondly, there was the lapse of time between the 4 June 1941 returns (or the last updating of the Parish Lists) and the date of the Primary Return. Then there were often changes in ownership and in acreage; holdings had been added and old ones had disappeared; and land had been requisitioned by the military, or taken over by the CWAECs. Thirdly, there was the problem of the definition of an amalgamated holding: it had been decided to exclude from the NFS all holdings of less than 5 acres. This meant in a large number of cases that holdings that were made up of an aggregate of individual dispersed holdings of less than 5 acres might have only part of their whole area recorded, or even be missed by the survey altogether. The result was that whole sections of the agricultural land of certain counties were going unrecorded. A reinterpretation of the 'less than 5 acres' rule by the Ministry of Agriculture, indicating that land of less than 5 acres that was farmed in conjunction with other land should be included in the survey, led to the need to obtain records for holdings that were individually less than 5 acres, and in each case to establish the association with the other blocks of land making up the whole. Differences of opinion in this by the five or six people involved in making up the farm record of each such holding meant that it was certain that uniformity of interpretation in this regard would never be obtained over the whole country, or even within districts and parishes.

Finally, there was the issue of the changing managerial classifications of the farms. The trend was towards creating 'A' farms, and a great deal of effort had been put in by the CWAECs to achieve the upgradings. Before June 1941 there had been very few dispossessions, but after that date the intolerance of the bad farmer by the CWAECs had been more marked. Hence, a delay between the 4 June Census Return and the Primary Return could well lead to farms being shown under new gradings that had not prevailed on 4 June 1941.

From this evidence, it can be seen that the completion of the farm record by the matching of the Primary Return with the Census Return could be a very difficult matter.[39] In March 1942, it was decided that if a Primary Return could not be matched with a Census Return because the latter had not been issued or returned, then it was too late to reinstate the June return and the Primary Record should be filed away but excluded from any future analysis.[40]

Problems with the maps

Circular no. 545 of 26 April 1941 had asked the CWAECs to include a map showing the boundaries of each farm as part of the complete farm record. Many CWAECs were very reluctant to do this work, owing to the time involved: they considered it an unnecessary addition to the Survey. CWAECs also felt that the 6-inch sheets would be too small a scale to show holdings of a 5-acre size and that the work should be done using sheets at the 25-inch scale.

The bombing of the Ordnance Survey's offices at Southampton in September 1941 resulted in the loss of the majority of stocks of 6-inch and 25-inch map sheets. The CWAECs were asked in November 1941 to send as a priority their requirements for 6-inch sheets (which might be also be obtainable 'from other sources') so that urgent reprints could be put in hand. To reprint 25-inch sheets, however, would be a much longer process, and the CWAECs were asked to locate their own copies of these sheets to be sent to the Ministry for photographic reduction. Further losses caused by enemy action took place in June 1943, when the Devon WAEC district offices in Plymouth were bombed and various of the 6-inch and $12\frac{1}{2}$-inch (reduced 25-inch) map sheets were lost 'along with other valuable records'. The Devon work had to start afresh.[41]

The view was expressed in February 1943 that the mapping was proving very troublesome owing to the fact that the Ordnance Survey sheets were not up to date and did not show recent developments.[42] In addition, the depiction of small holdings was proving extremely difficult, and they presented such a complex picture that it was felt that they could serve little useful purpose. The Ministry's Planning Branch, however, disagreed and stated that in and around urban areas the 'complicated mosaic' was exactly what was required. The Ministry of Agriculture emphasized that the mapping must be continued, and that all holdings of 5 acres and more must be shown.

At the meeting of the Farm Survey Supervisory Committee held on 12 July 1943, the point was made that many of the farm boundaries shown on the maps were already out of date as the speed of change was so great. It was felt that the mapping should be postponed until after the war. However, a decision was made that the farm boundary mapping should continue, because, however imperfect the maps in some respects, they would nevertheless be invaluable for future planning to help to prevent development from cutting across the boundaries of efficient farm units.

In September 1943, the Ministry of Agriculture had instructed that the NFS maps be dated. Some CWAECs, however, dated their maps '31st March 1942', which was the date by which the survey should have been completed rather than that which was required by the Ministry for the gathering of the information on the farm boundaries. Again, the Ministry's instructions had not been specific enough. Doubtless, some CWAECs used other 'convenient dates', probably including that of the completion of the map, which might be months, or even years, later than the determining of the farm areas.

Problems with finance

The total grant from the Development Fund to the NFS in its first year was £20,000. By February 1942, the anticipated cost had risen to £145,000, which brought a sharp letter of rebuke from the Treasury at the incomplete picture of the costs that had been presented to them.[43] A large part of the extra costs was made up by payments to CWAECs for additional staff, for the purchase of maps (£7000–£8000), the hire of tabulating machines (£6000), payments to Crop Reporters who had the task of clearing up errors on the page 3 supplementary form (£5000), and in the fact that the Survey went on for at least 2 years longer than anticipated, with the consequent need to keep in place the administrative arrangement set out and costed above for the first year only. However, owing to the economies found in the mechanical tabulation process, as described above, it is likely that the total expense was short of the £145,000 figure anticipated in 1942 – perhaps £140,000 (some £7 million in late 1990s terms).[44]

This cost was calculated at 9s 9d per holding, which was said to be very good value for money in the light of the information obtained for each holding and the map drawn. The sum of 9s 9d contrasts with a figure of £1 a holding that would be a land agent's professional fees for similar work. Emphasis was placed on the voluntary, unpaid work undertaken for the survey by many individuals involved with the CWAECs. This had kept down the cost considerably.[45]

Archival Policies for the National Farm Survey

Confidentiality

From 1943 until the mid-1950s (the time of the transfer of the farm records, excepting the maps, to the Public Record Office (PRO)) the MAF received many requests for access to the records of the NFS.[46] Indeed, paragraph 14 of the Summary Report had stated: 'Requests for access to this material for research purposes will be sympathetically considered, so far as is compatible with fulfilment of the undertaking that all particulars relating to individual farms will be regarded as confidential.' Some of the requests came from students and other researchers, while others were of an official or semi-official nature.[47] Generally, it was easier to gain permission to scrutinize the maps, which were considered not to contain information obtained in confidence, than the written farm records. It seems that at no time was direct access to these records allowed to anyone who was not acting in an official capacity. The most that the Ministry would allow was the extraction of certain data by the Advisory Centres, but always in the form of an aggregate spread over several parishes, so that no individual holding could be identified.

The Ministry laid down firm rules to determine which official or semi-official applications for direct access to the farm records might be allowed, as well as instructions for the borrowing of NFS material if permission actually had been granted. Borrowers were asked that 'strict precautions be applied against risks of soiling, tearing or loss, and that each document be regarded as a valuable historical manuscript, which in due course it will become'. They were also reminded that 'the farm records contain a good deal of information of a confidential character, such that disclosure would involve a breach of faith or even give rise to an action for libel. Prosecution under Defence Regulation 84 would also be a possibility.' It was stated that all material when not in use must be kept in a locked and fire-proof cabinet, and that under no circumstances must it be taken into the field or otherwise exposed to the weather.[48]

Requests for information contained in the NFS include those from county council planning departments, and from various electricity companies and water boards, asking for details from the records to assist their post-war reconstruction and development.[49] Some information might be allowed by the Ministry and extracted from the farm records by the Advisory Centres, but free access to the records was never granted for such requests and certainly not to any of the sensitive data relating to management. In most cases, the answer was a firm negative:

> We have been trying very hard to see if we could help you with the information you want. But I am afraid at the moment it is hopeless. In a word the position is that we have got a great deal of this information on the condition that it is treated as confidential and we have got to be almost unbelievably careful. It's just like the Inland Revenue and the individual's private income. So sorry.[50]

Favourable consideration, however, to obtaining details from the farm records was granted to bodies such as the Country Landowners' Association, wishing to compile lists of the names and addresses of agricultural landowners. Dudley Stamp, in his capacity as Chief Adviser on Rural Land Use to the Minister of Agriculture and Fisheries, was also allowed permission to borrow farm records to carry out test surveys of typical rural areas – permission that was extended to certain of his staff. An application by C.S. Orwin of the Agricultural Economics Research Institute at Oxford to see some of the farm records, in connection with the economic and social survey he was carrying out, was at first denied but later he was allowed to obtain extracts from the records.[51] Letters from solicitors for the Duke of Bedford's Tavistock Estate requesting details of the grading of farms on the estate, and from Earl Bathurst's Hunt asking for permission to see the maps showing farm boundaries so that they would know whom to contact if they caused any damage to fences and hedges, received quick and decisive rejections.

Restrictions imposed by the confidentiality of the information on the management grading do seem to have prevented the records being used by planning and other bodies as fully as might have been desired: the result was thus the opposite of what had been announced as one of the main reasons for the NFS. The MAF was aware that, having spent so much money on the survey, its use should be as wide as possible, but they were caught in a dilemma by the confidentiality of the information that they had obtained.

In 1949 officials at the Ministry sought to re-establish under what conditions the survey had been carried out, and what exact undertaking had been given to farmers regarding the confidentiality of the information that they had been asked to supply. This does not seem to have been exactly recorded at the time, but the fact that field officers did reassure farmers on this point was not doubted by the Ministry. Reliance on this matter was placed in Defence Regulation 55(1)(d), which was to be read with

Regulation 84, which dealt with the non-disclosure of information obtained under the Defence Regulations. All the files and correspondence show that the Ministry – and the Advisory Economists – took their responsibilities and obligations to the farmers very seriously, and there is no recorded instance where that trust was compromised or breached.[52]

The transfer of the archive

The NFS was seen almost from its inception as a major historical record – a second Domesday Book – that would be preserved in perpetuity as part of the national archive. Its value for future researchers was perceived at an early stage in its compilation. Great care was therefore taken to preserve it in its entirety, and to prevent loss and damage to the various forms and maps making up the total farm records. Unlike the maps, within a few years of the end of the war the written farm records soon ceased to have any working use for the MAF, and their remaining value resided almost solely in their 'Second Domesday' status.

Between January 1957 and June 1959 the Ministry's Archive Section, working at Hayes, recalled the Primary Returns and the Census Returns (but not the maps) in batches from the Advisory Centres and packed, referenced, listed and boxed these meticulously in the form in which they are produced by the PRO today. However, the records were not assembled 'matched' (i.e. the Primary Return placed with the Census Return forms for each holding), as had presumably been the arrangement in the dossiers at the Advisory Centres. The five forms making up the farm records for each parish were now arranged in groups, with all the Primary Returns together, and all the Census Return forms in their individual groups, each arranged in numerical order of the farm code references, laid side by side within envelopes for each parish. These individual farm records were given the archival class reference MAF 32.

Of the dossiers maintained by the Advisory Centres on each farm within the 12,000 parishes of England and Wales, it was just the Primary Records and the 4 June 1941 returns that were transferred back to the Ministry. All the other material that the Advisory Centres had included in the dossiers, including for certain counties additional survey information and individual farm plans, appears to have remained with the Advisory Economists. The first Advisory Centre to transfer its records in this way was the Western Province at Bristol, which sent to the Ministry's Hayes repository the records for Gloucestershire, Herefordshire, Somerset, Wiltshire and Worcestershire. The other Advisory Centres followed in due course.[53]

Once the records had left the Advisory Centres for the Hayes repository, there to be packed for their eventual transfer to the PRO, access to them to obtain edited extracts for research purposes was not so readily allowed. The

Ministry's Archive Section did do some work for students, but this took so long that the Ministry felt it could no longer justify the time and expense that the work entailed, although it still continued to look at student applications on their merit. Once the records were in the custody of the PRO, access to them was not possible under any circumstances, unless the Ministry itself wished to recall the records to carry out work on them. There is nothing in the correspondence to show that this was ever done.

The PRO had, of course, been consulted about the eventual transfer to it of the records of the NFS, and it agreed to receive the written farm records separately ahead of the maps, which were still in use. The complete MAF 32 archive consisted of the records of some 320,000 holdings, which took up about 180 m of shelf space. Because the NFS had been conducted under a promise of confidentiality to the farmers, a closure period of 50 years was agreed. The date from which the 50 years was to be calculated was taken as 1941, which meant that the records would be available for public scrutiny on 1 January 1992.[54]

In 1959, the MAF 32 records were physically transferred to the PRO at Chancery Lane, where they were locked in strongrooms because of their 'extended closure' status. They were sent to the PRO's new building at Kew in 1976/77, where again they were placed in strongrooms. January 1992 saw their due release for public scrutiny.

In 1961 the Parish Lists for 1941, which form an essential part of the NFS archive, were transferred to the PRO.[55] A decision had been taken by the MAF to preserve these because of their direct relevance to the NFS. The normal practice was to destroy the Parish Lists after 5 years, although in 1961 those for 1939–1953 still survived. In 1961, a decision was also taken to preserve the Parish Lists for 1960, as these tied in with a Farm Management Survey (based on the 1960 Agricultural Returns) carried out by the Ministry. The Ministry of Agriculture, Fisheries and Food's Legal Department considers that the provisions of Section 80 of the Agricultural Act 1947 apply to the Parish Lists, which means that they cannot be examined without the written consent of the Ministry. Various attempts over the years have been made by the PRO to have this restriction amended to a straightforward closure period, but without success.

In 1948 the MAF decided that the NFS maps should be transferred from the Advisory Economists to the Provincial Land Commissioners. This caused a storm of protest from the Advisory Economists, who felt that they would be severely inconvenienced as they needed frequent access to the maps for various projects in which they were involved. A concern was also expressed by the PRO (which had been consulted) about possible damage or defacement to the maps by their further use. The matter was resolved by allowing the Advisory Economists continuing access to the maps, including the right to borrow them. The PRO agreed to the new arrangement when they were assured that the maps would be used 'in such a way as would not

appreciably impair their subsequent value to the PRO'. On no account
would they be used in the open as 'field sheets'.

The Provincial Land Commissioners were asked to deal with the
Advisory Economists very tactfully: 'These Economists have put a good
deal of work into the maps and their colleagues have housed all the farm
survey material, which takes up much space, for several years.' This tact
was also expressed by the Ministry. A letter from Kirk to Wye College
states: 'We shall not have to ask you to give up the custody of the maps in
your care until the time comes when they have outlived their usefulness as
any but historical documents and are handed over to the Public Record
Office.' A memorandum to the Provincial Land Commissioners dated
September 1948 also asked that, where there were satisfactory copies, they
consider allowing the Advisory Economists to retain the originals. In fact,
by 1948 the great majority of the NFS maps had been copied on to $2\frac{1}{2}$-inch
sheets and most Advisory Centres were able either to retain the originals or
to obtain a $2\frac{1}{2}$-inch copy.

In 1956, when the transfer of the farm records to the PRO was first
being discussed, it was decided that, as the maps were still being used by
the Provincial Land Commissioners, the transfer of the farm records should
go ahead without their accompanying maps. In 1958, however, the Land
Commissioners surrendered the original maps to the regional offices of the
MAF. They were transferred from the Ministry's regional offices to the PRO
Hayes storage depot in 1964. Here they were packed in special manila folders, then boxed (which is the form in which they are produced today) and
listed within the document class, with the archival reference MAF 73.[56]

By September 1966, the packing and boxing process was nearly complete, with 3621 folders containing 36,000 maps. The packing and listing had
shown where map sheets were missing, either because they were never
completed or because they had been lost. These are marked in the class list
as 'wanting', and the diagrams stamped on each map folder also show where
sheets are missing. The only really substantial blocks of maps 'wanting' are
in Durham and Monmouthshire, together with the Isles of Scilly (Fig. 4.2).

There was a delay from 1966 to 1970 in the physical transfer of the
maps to the PRO at Chancery Lane owing to a dispute over their period of
closure. The PRO had queried whether the 50-year closure period to which
the farm records in MAF 32 were subject had also to be applied to the
maps. The Ministry of Agriculture confirmed that the 50-year closure was
equally binding on the maps, and also stated that the operative date should
be 1943, rather than 1941 (which had been accepted for the farm records).
Faced with the prospect of opening the individual farm records in 1992 and
the maps in 1994 – even though both were part of the same total farm
record – the PRO suggested a compromise by which the maps should only
be closed for 48 years, thereby giving them a 1992 opening date. The
Ministry agreed to this.

The physical transfer of the maps to the PRO at Chancery Lane took place in 1970. They were also moved to Kew when the PRO's new office opened in 1977, and duly became available for public scrutiny on 1 January 1992.

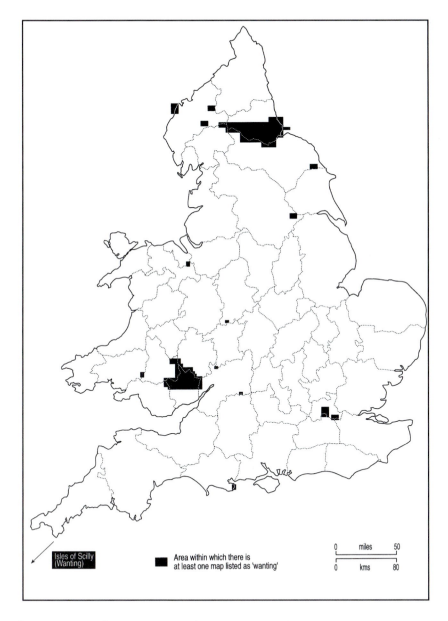

Isles of Scilly
(Wanting)

■ Area within which there is
at least one map listed as 'wanting'

0   miles   50
0   kms   80

Fig. 4.2. Areas with missing National Farm Survey (NFS) maps.

Preservation and destruction

The Ministry of Agriculture reviewed methodically their wartime records, in consultation with the PRO. Such reviews included their internal policy and administrative records as well as those of their external agencies, such as the CWAECs. When the various periods of retention were complete the records would have been subject to a second review, and a decision may well have been taken to preserve a few indefinitely for eventual transfer to the PRO. Certainly the files in PRO MAF 38 relating to the administration and analysis of the NFS were preserved, but many others were destroyed. Where the records of the CWAECs were to be destroyed the procedure appears to have entailed taking the material to the local Head Postmaster to be treated as 'waste' and to be pulped and recycled, but various committee members seized the opportunity to retain papers relating to their own work with the CWAECs and District Committees. The extant county record office papers such as the Wigley Papers at Buckinghamshire CRO or the Leigh estate papers in the St Helens Local History and Archives Centre undoubtedly survived in this way, and others may yet survive among estate archives.

Curiously, some CWAEC wartime records survive in administrative use to this day. The Ministry's East Midlands regional office holds files on individual farms in Derbyshire, Huntingdonshire and Lincolnshire which had been started during the war, and which include copies of the NFS Primary Return. Certainly many other regional offices hold such information, since many CWAEC (later CAEC) files were passed on to the Ministry's regional officers when they came into post from 1957. The CAECs were themselves finally closed in 1972, the two systems operating side by side for 15 years but with liaison and presumably some transfer of records from one to the other. Correspondence between the Ministry and the Advisory Economists relating to the NFS and records of the conferences of Advisory Economists are held at the PRO, together with other records of the Advisory Economists.[57]

Checks on the individual farm records (PRO MAF 32) show these to be substantially complete for England and Wales, although a few parishes have so far been found as having no papers. For Warwickshire, additional parish records were found at a later date at the Reading Advisory Centre and these were added to the list in the PRO at a later date. One unresolved problem is the location of the records for the Isles of Scilly, which formed part of the South Western Advisory Province. It is also not yet clear whether the Isle of Man was actually included in the NFS. Finally, although this particular volume concentrates on the NFS records and their production, it is important to appreciate the context of the records within the large body of other wartime material available at the PRO and elsewhere. The wartime records of the Ministry are also in many cases continuations of the pre-war classes (Cantwell, 1993: 28–32).

Notes

1 PRO MAF 38/205; 209 (FSC10, 11, 12).
2 PRO MAF 38/209 (FSC11).
3 PRO MAF 38/215.
4 PRO MAF 38/205 – this file contains detailed correspondence on the problems of the completion of the records and bringing the survey generally to a state where it was ready for analysis.
5 PRO MAF 38/209 (FSC14).
6 PRO MAF 38/209 (RSC3 and RSC4).
7 PRO MAF 38/209 (RSC4). The Summary Report of 1946 reports fully on the sampling procedures, drawing attention to the definition of the population and the drawing of the sample, as well as the adequacy of the sample. See the report, Appendix I, 71–77.
8 PRO MAF 38/207 (FSC15).
9 PRO MAF 38/297 includes a copy of the Preliminary Instructions for Coding issued in November 1942 (those used for the Kent pilot sample). The Final Instructions do not appear to survive amongst records held at the PRO. PRO MAF 38/297 also has a copy of a coding sheet, but again this presumably relates to the preliminary instructions.
10 PRO MAF 38/205.
11 PRO MAF 38/205.
12 The research programmes are described at length in PRO MAF 38/472.
13 PRO MAF 38/472 (EF541).
14 PRO MAF 38/207 (TDY3637).
15 PRO MAF 38/865.
16 PRO MAF 38/209 (RSC6).
17 The county appendices (or county results) are in PRO MAF 38/852–863. They are particularly interesting, for they show the way the data was organized for the tabulation process, and the types of cross-tabulations of the different data items that were possible.
18 A brief summary of the final report was written and published by H. Whitby, 'National Farm Survey of England and Wales', *Agriculture* 53 (8) 1946, 335–340.
19 PRO MAF 38/473. The material that was cut from the draft comprised paragraph nos 89–90 and Table 34.
20 PRO MAF 38/216.
21 A proof copy of the report is in PRO MAF 38/216.
22 PRO MAF 38/216 (EF569); and see the *Summary Report* Preface by Williams.
23 PRO MAF 38/474.
24 PRO MAF 38/475 and PRO MAF 38/695.
25 PRO MAF 38/205.
26 PRO MAF 38/207(52).
27 PRO MAF 38/214.
28 PRO MAF 38/256.
29 Various other suggestions for the easier and cheaper resolution of the page 3 queries are set out in a memorandum of 4 March 1942 – PRO MAF 38/207 (42).
30 PRO MAF 38/209 (FSC9); 38/214.
31 PRO MAF 38/209 (FSC9).
32 PRO MAF 38/471 includes a letter expressing this concern from the Northern District Committee of the Hampshire WAEC.
33 A.G. Street (1892–1966), farmer, author and journalist, did not go to gaol, although his Primary Return was transcribed from L.G. Huntley's surveyor's report as Q.3 'Farmer makes definite refusal to disclose this information, as a matter of principle'. Street's Ditchampton Farm in Wilton, Wiltshire, was classified as an 'A' farm (PRO MAF 32 50/235).
34 G.H. Garrad later wrote *A Survey of the Agriculture of Kent* (1954).
35 PRO MAF 38/205.
36 PRO MAF 38/205. See also F.W. Bateson (ed.) (1946) *Towards a Socialist Agriculture: Studies by a Group of Fabians* (Gollancz) in which he draws heavily upon his Buckinghamshire work with the CWAEC. Frederick Wilse Bateman (1901–1978) was a Fellow and Tutor in English Literature at Corpus Christi College, Oxford 1946–1969 and

was editor of the *Cambridge Bibliography of English Literature* 1939–1940. He had been a WEA lecturer (1935–1940) and became Statistical Officer for the Buckinghamshire WAEC (1940–1946). Publishing *Mixed Farming and Muddled Thinking* in 1940, he was agricultural correspondent for *The Observer* and *New Statesman* (1944–1948), and a keen Buckinghamshire local historian (see his *Brill: a Short History* (1966)).

37 PRO MAF 38/205.

38 PRO MAF 38/214 and PRO MAF 38/205, in particular, are full of correspondence between the Ministry of Agriculture and the Advisory Centres on the very detailed queries arising out of the survey, and on the matching process.

39 A detailed memorandum by S.R. Wragg of Harper-Adams, with comments from the Ministry of Agriculture, sets out further the problems of matching the Primary Returns with the Census Returns – see PRO MAF 38/212 (8). In addition, Ministry correspondence with A.C. Baker and his principal, J. Wyllie, at Wye throws much light on the difficulties of matching the records and bringing the survey to a cohesive conclusion – see PRO MAF 38/205.

40 PRO MAF 38/207 (42).

41 PRO MAF 38/865 (74–78).

42 PRO MAF 38/212 (17).

43 PRO MAF 38/470.

44 PRO MAF 38/215.

45 The voluntary nature of much of the CWAEC activities during the war has been frequently commented upon. Details of the costs of the Advisory Centres in the NFS are given above, as also is the expenditure on the Crop Reporters in their role of clearing up the page 3 queries.

46 PRO MAF 38/865 and PRO MAF 38/866.

47 The files are full of research proposals that might now, with the public availability of the NFS records, be carried out in full.

48 PRO MAF 38/865 (35).

49 PRO MAF 38/865. Further requests from electricity companies are in PRO MAF 38/471. Requests regarding water supplies are in PRO MAF 38/474.

50 PRO MAF 38/471.

51 PRO MAF 38/471. C.S. Orwin (1876–1955) was the Director of the Oxford Agricultural Economics Research Institute from 1913 to 1945. He was a Fellow and Estates Bursar at Balliol College (1926–1947) and had been Editor of the *Journal of the Royal Agricultural Society of England* (1912–1927). There seems to have been a measure of ill-feeling by certain Advisory Economists concerning Stamp's perceived privileged role in so many projects concerning the NFS.

52 PRO MAF 38/865. See also PRO MAF 38/471 (EF214).

53 It was the order in which the records were received at Hayes that dictated the order in which they were listed. This explains why the PRO MAF 32 list is not arranged alphabetically by county.

54 As the NFS is generally dated between 1941 and 1943 (although some individual records are dated 1944, and even 1945), a more consistent opening date would have been January 1994.

55 PRO MAF 65.

56 The Ministry must also have recalled any originals that were still with the Advisory Centres as there are no sheets of the 2½-inch copies in PRO MAF 73.

57 PRO MAF 38/198–204; 205–217; 466–476; 852–867.

An Assessment of the National Farm Survey Data
5

In this chapter we provide a detailed assessment of the range and quality of the data provided by the National Farm Survey (NFS). We consider the nature of the responses required on the forms, how entries actually varied, and how they were either misunderstood, mistaken, amplified or subverted by the surveyors, copiers and farmers who were responsible for completing the forms. We also assess the internal consistency of the data to gain an understanding of its quality and potential uses.

The assessment is based on the examination of 1450 holdings from the *National Sample*, together with 1200 holdings in the *Sussex Sample* and 480 in the *Midlands Sample*. This provides a sense of national, regional and local variation in the quality of the data. The amount of information provided in this chapter for each item of data varies considerably. For some variables, such as **County**, there is no need for an extended commentary; for others, such as **Code No.**, a fuller description is required. The many items which required a straightforward numerical response are not greatly dwelt upon.

The data discussed in this chapter are found on four different forms within the Ministry of Agriculture and Fisheries (MAF) 32 in the Public Record Office (PRO):

1. The *Primary Return* or Primary Farm Record [B496/EI].
2. The 4 June 1941 Census Return (Crops and Grass, Labour, Livestock and Horses), called throughout the *Census Return* [C47/SSY, p. 1].
3. The 4 June 1941 Census Return (Small Fruit and Horticultural Produce), called throughout the *Horticultural Return* [C51/SSY, p. 2].

4. The 4 June 1941 Census Return ('Supplementary Form' concerning Motive Power, Rent and Length of Occupation), called throughout the *Supplementary Form* [398/SS, p. 3].

It may be useful to refer back to Figs 1.1–1.4, which show the layout of the forms and illustrate how the design of the forms and the amount of space allotted for each answer placed certain restrictions upon those recording the data. All four forms survive for over three-quarters of the holdings but a significant minority of holdings have missing forms. Table 5.1 shows the number of forms in the three samples. In the *National Sample* 85.4% of the holdings had four forms; in the *Sussex Sample* it was 80.0%; in the *Midlands Sample* 78.2%. The form most likely to be missing was the Primary Return. Table 5.2 is a summary of the *National Sample* by Advisory Centre provinces. The proportion of holdings with all four forms varied from 78.2% in South Eastern province to 100% in North Western, with most being around 80–85%.

In this chapter the data entries are described in the order in which they are found on the various forms that make up the NFS. They are stored, parish by parish, in PRO MAF 32.

The Primary Return

[ACREAGE]
Extraordinarily, no place for the total acreage of holding was designed into the printed Primary Return form. The County War Agricultural Executive Committees (CWAECs) were instructed by MAF to insert a figure for the total acreage at the top right-hand corner of the Primary Record. This was most often done by hand, sometimes in a space marked out by some kind of

Table 5.1. Numbers of the various forms processed into the databases.

	Name of database		
	National Sample	Sussex Sample	Midlands Sample
Number of records processed	1450	1200	480
Number of holdings with 5+ acres	1330	1186	477
Primary Return (B496/EI)	1238	1056	416
4 June Return (C47/SSY, p. 1)	1402	1161	454
Horticultural Return (C51/SSY, p. 2)	1399	1154	445
Supplementary Form (398/SS, p. 3)	1389	1146	453
Number of holdings with all forms	1136	949	373
Percentage of total records processed	78.3	79.1	77.7
Percentage of holdings with 5+ acres	85.4	80.0	78.2

Table 5.2. Numbers of the various forms processed into the National Sample Database, by Advisory Centre Provinces.

Advisory Centre Province	Number of records processed	Number of holdings with 5+ acres	Primary Return (B496/EI)	4 June Return (C47/SSY, p. 1)	Horticultural Return (C51/SSY, p. 2)	Supplementary Form (398/SS, p. 3)	Number of holdings with all forms	Percentage of total records processed	Percentage of holdings with 5+ acres
East Midland	130	123	115	127	124	120	100	76.9	81.3
Eastern	173	173	167	170	168	167	157	90.8	90.8
North Western	45	31	31	37	45	37	31	68.9	100.0
Northern	89	89	81	88	86	86	79	88.8	88.8
South Eastern	133	133	115	125	127	128	104	78.2	78.2
South Western	71	64	57	68	70	70	54	76.1	84.4
Southern	182	135	120	178	182	182	115	63.2	85.2
West Midland	67	67	66	58	62	61	54	80.6	80.6
Western	103	98	86	100	94	97	78	75.7	79.6
Yorkshire	105	97	95	104	104	103	93	88.6	95.9
Wales:	352	320	305	347	337	338	271	77.0	84.7
North Wales	214	184	169	213	206	208	149	69.6	81.0
South Wales	138	136	136	134	131	130	122	88.4	89.7
National Sample	1450	1330	1238	1402	1399	1389	1136	78.3	85.4

stamp. The reason for the addition to the Primary Return of the acreage figure was to assist in the 'matching' process of Primary Records and Census Returns at the Advisory Centres. The figure may be termed 'Acreage', 'Total Acreage' or 'TA' and may be accompanied by figures for 'Crops and Grass' and 'Rough Grazing', or even for individual crops, but this was not done consistently at a regional or national scale. Within the *National Sample*, of the 1330 holdings that had a Primary Return only 61 (4.6%) did not give a figure for acreage. In the *Sussex Sample*, of the 1186 holdings that had a Primary Return only 7 (0.6%) did not give a figure for acreage; while in the *Midlands Sample* over one-fifth (100; 21%) of the 477 holdings that had a Primary Return had no acreage figure. More information about the quality of acreage information is provided in the section on farm size and structure in Chapter 7.

[TITLE]
Some West Sussex records of market garden holdings have the 'Farm Survey' title altered to 'Horticultural Return'. This probably relates to a horticultural survey that the CWAEC was carrying out at the same time as the NFS, the records of which became (at least in part) incorporated with those of the NFS.

COUNTY
This should be a straightforward recording of the county name. However, there can be many variations in the form in which the name is given (e.g. 'Bedford' and 'Bedfordshire') or it is given in an abbreviated form (e.g. 'Cambs' or 'Soke of P'). Certain East and West Sussex parishes are not designated as such, but 'Sussex' alone is recorded.

CODE NO.
The CWAECs were instructed by MAF to use the farm reference system already established for the quarterly census statistics. This was generally done in the *Sussex Sample* but occasionally another code system was introduced into the reference, which can cause confusion. An example is XE/196/57/15 instead of XE/330/57/15, where either a Crop Reporter's code (196 instead of 330) has been used, or where there has been some confusion over the area administered by that Crop Reporter. In some instances a code number that is included into the reference appears to relate to a numbering system used by the CWAECs for the districts within their counties, rather than the MAF system of numbered Crop Reporters' areas.

Where a holding consisted of several parts, each having separate codes for quarterly census purposes, these were usually added against the holding's principal code number – for example, XE/11/148/68 and 41 and 77. Usually there will not be separate Primary Returns for these additional holdings, but they will almost certainly have separate Census Returns. The additional code numbers are normally from within the same parish, but occasionally they are from other parishes in cases where the fragmented holding spreads over a wide area. If a holding had not been making quarterly Census Returns there was no code to apply to it. The Primary Returns for such holdings have the 'Code no.' space left blank or some alternative code in its place.

In the *Midlands Sample* there were similar variations in the code numbers. In Nottinghamshire they appeared in the format 'NOT/5/23/1'. In Hereford it was specified as a different type of code: 'County Code No. 18/1'. In Derbyshire it was in the format 'DER/1/[some number, e.g. 32]' for all holdings. The first part of the code here appears to have been written in beforehand for the copiers/surveyors to fill in, but they later reverted to the farm reference code.

Within the *National Sample* it was normal for the farm reference code specified by MAF to be used. However, it was not rare for other codes to appear on the Primary Return and on the addressograph side of the Census Return and its data side. These codes could be variations on the components of the farm reference code (as in Bedfordshire), a simple numerical code (again as in Bedfordshire), or some coding system used by the CWAECs comprising letters and numbers (as in Derbyshire). In Leicestershire the code appeared on the Primary Returns and maps in the format 'LR/AS/439', the letters referring to the County and the District (Ashby de la Zouch). Sometimes the farm code reference was given in part only, e.g. 158/32 (just the parish and the schedule numbers). Rarely, but certainly noted in one county, just the schedule number (the farm number) is given. There are also codes that do not appear consistently and are not connected to other documents, e.g. at Stackpole Elidor (Pembrokeshire). In Glamorgan the farm reference code varied between the documents, starting 'GM' on the Primary Record but 'GN' elsewhere. In Essex, as well as the farm reference code on the Primary Record there is the Essex WAEC code on the Census Return. There is also another code, which appears next to the 'Management' section of the Primary Record but does not correspond with a similar code on the Census Return. In Cornwall a variety of colours were given as an additional code, e.g. 'Yellow' and 'Violet'. The first part of the farm reference code was sometimes stamped on, with only the last two numbers to be filled in by hand on the form. Sometimes a stamp was applied to the top of the Primary Record, with space for the acreage figure (see below) and the Code No. to be inserted. Where this happens the county and Crop Reporter's codes were inserted in the printed 'Code No.' space on the Primary Record and the parish and farm codes were added within the appropriate space within the stamp.

DISTRICT

The name of the CWAEC district, which usually corresponds with the Rural Districts, but not necessarily with the Crop Reporters' areas, was entered here. Sometimes the name was stamped. Occasionally there can be some confusion as to the exact name of the district (e.g. Worthing, or

Chanctonbury and Worthing). Rarely, when the land area of the holding extended into another district, or when an amalgamated holding consisted of parcels of land in more than one district, more than one district would be entered. Designations other than place names were sometimes used – such as in Essex, where the number 3 or 'Chelmsford 3' were used; or in Soke of Peterborough, where the term 'Centre' was used; in Warwickshire, where 'E2 – Stratford' was used; in West Suffolk, where 'No. 5' was used; or in Westmorland, where 'West Ward' was used.

PARISH

The name of the parish was entered here. As with District, if a holding extended over more than one parish, or if it had separate parcels of land in other parishes, then the names of the other parishes were often entered as well. Spelling could vary widely; see for instance Mansel(l) Lac(e)y, Herefordshire.

NAME OF HOLDING

This was usually just a straightforward entering of the name of the farm or other holding. Quite often, however, the name was different, either totally or in detail of its spelling, from that given on the Census Return forms. Often, the addressograph plate (from the Parish Lists) was out of date, but the error may result from the surveyors or the copiers of the records. It seems that certain farms had no definite form of the spelling of their names, and there might be two or more variants. The Horticultural Return can often provide a good check on address details, as it was copied directly from the farmers' own entries of their addresses.

Variations in the manner in which the name of the holding was entered are as follows:

- Often just a name was given, without 'Farm' being added.
- Sometimes the holding had no name, or it was described as, for example, 'Land at ...' or 'Land south of ...' or simply 'Land'.
- Sometimes the name was entered as, for example, 'Holly Farm (part)', and there was likely to be a further record for that farm elsewhere within the parish.
- Where a holding was made up of several farms, sometimes the names of all the farms were given (with presumably the first being the principal – although this does not always tie up with the information on the Census Return forms), but at other times just the name of the principal farm.
- Where an owner, or tenant, did not actually live on the holding, there was sometimes confusion between the name of his residence and the name of the holding; for example, a farmer living at Rose Cottage, Denton, but farming a piece of land a mile away, may have the name of his holding entered as Rose Cottage.

NAME OF FARMER

As with 'Name of Holding', this entry should tie up with the information given on the Census Return forms. Usually, just the initials and surname of the farmer were given, but in some counties full Christian names were recorded. Often, the Horticultural Return (see 'Name of Holding' above) provides details of Christian names. The entry normally recorded titles (other than Mr), and if the farmer was a Mrs or Miss this was normally stated. Other titles frequently met are Lord and Lady, The Hon., Col., Major, Brig. Gen., and so on. If the farm was owned by more than one individual, this should be entered accordingly, for example 'R. & T. Edwards', as is the case if it was run as a business, for example 'Messrs. P.T. Cox & Son'.

Where the name on the Primary Record does not match with that on the Census Return, this was usually because there had been a change of owner between the 4 June 1941 return and the date the survey was carried out (sometimes 2 years later, or even longer). Often the initials on the Primary Record will be different from those on the Census Return, but the surname the same: this is the case where the farm had passed to, say, a son or to a wife, especially in the latter case because the husband was away on active service. In cases where the land had been taken over by the CWAEC, the name of the Executive Officer of the CWAEC was often entered as the farmer.

ADDRESS OF FARMER

Very often this entry was simply a repeat of the information given in 'Name of Holding' and 'Parish'. However, where the farmer did not actually live on his holding, then his home address was given here (see 'Name of Holding' above). The addresses should be the same as those given on the Census Returns, and in the case of farms in small and remote villages they may also have provided the name of the nearest major town.

NUMBER AND EDITION OF 6-INCH ORDNANCE SURVEY SHEET CONTAINING FARMSTEAD

In its fullest form the entry here will provide the reference of the relevant Ordnance Survey (OS) sheet(s) (using usually arabic numerals, but sometimes roman) at either the 1:2500 or 1:10,560 scales, and the date of the edition. Often, however, the date was not given, and very often (although not the case in Sussex) the county title was not added. Although the instruction was for the sheet reference to be given where the farmstead was situated, often references were provided to all sheets over which the land area of the holding extended. Many holdings in any event did not possess a farmstead. The scale of the map sheet given here is not necessarily that of the sheet actually used to provide a record of the farm boundaries. Sometimes, as in Bedfordshire, no information was given at all.

Section A: Tenure

IS OCCUPIER TENANT OR OWNER?

This question, which demanded the placing of a cross in the appropriate box, was sometimes answered by a cross in both boxes, i.e. the occupier was both tenant and owner of different parts of the holding. Often, in cases like this, the relevant acreages were written in – for example, 'Tenant (35 acres), Owner (12 acres)'. Table 5.3 shows the proportion of holdings that were owner-occupied, tenanted or mixed tenure for the three samples and indicates that well over a half of all holdings were tenanted at this time. Where land was both owned and rented the entries can be cross-referenced with the figures given on the Supplementary Form for the proportion of the holding that was owned as opposed to rented. This is discussed in Chapter 7 in the section on farm size and structure.

IF TENANT, NAME AND ADDRESS OF OWNER

The name and address of the owner (or of several owners) was written here. The addresses were sometimes in full, or in an abbreviated form (e.g. just the name of the town or village), or none was given. In the cases of some landowners, the name and address of their agents are given. There are some examples of the details of owners being so lengthy that they were carried to the 'General Comments' box on the rear of the form – such as in the Isle of Ely, where owners' names and addresses and the OS field numbers of the fields they owned were also given. Where there were several owners, the acreage owned by each was often provided. See Chapter 7 for a fuller discussion of land ownership.

IS FARMER FULL TIME FARMER/PART TIME FARMER/SPARE TIME FARMER/HOBBY FARMER/OTHER TYPE?

The relevant category was indicated by a cross. A further class, 'Professional Farmer', was introduced fairly late in the survey, and this was written into the form. Over two-thirds of the holdings in the three samples were classed as full-time (Table 5.4).

Table 5.3. Tenure of all holders.

Name of database	All holders with tenure details available	Owner-occupiers	Tenants	Tenant/owner-occupiers
National Sample	1226	340 (27.7)	763 (62.2)	123 (10.0)
Sussex Sample	1055	351 (33.3)	616 (58.4)	88 (8.3)
Midlands Sample	410	93 (22.7)	270 (66.0)	47 (11.5)

Percentages in parentheses.

OTHER OCCUPATION, IF ANY

If the farmer was not a full-time farmer, an entry may be made here – for example 'Publican', 'Quarryman', 'Farm labourer', etc. Sometimes a comment such as 'Retired' or 'Not a farm, but a racing stables' would be entered here. If the farmer was undertaking military duties (for example 'Coast watching', 'Captain, Home Guard', 'Squadron Leader RAF', 'War reserve constable') this information might be entered here as well. In Runcton Holme, Norfolk, a farm was held and owned by Evelyn Gee, who was also down as the field recorder or primary completer on many of the forms for this parish.[1] His other occupation was given as 'District Officer, WAEC'. This is one of the clearest examples in the sample of farmers surveying their neighbours' and even their own properties.

In the *National Sample*, 290 (23.7%) holders were recorded as having other occupations; the equivalent figure for the *Sussex Sample* was 233 (22.1%) and for the *Midlands Sample* it was 141 (33.4%). Table 5.4 shows the proportion of those holders with other occupations that were classed as full time, part time and so forth. A very wide range of other occupations was given. The full-time farmers with other occupations were mostly in activities associated with agriculture, such as 'agricultural labourer', 'cattle dealer', 'butcher' and 'milk retailer'. Where there was a large estate in or near the parish, other occupations would reflect employment on the estate, such as 'estate carpenter'. The less full time the farmers were, the less their other occupations were associated with agriculture, although they might be given as landowners and there were some professions associated with agricultural research, such as 'Zoologist at Essex Agricultural Institute', 'Professor of Agriculture' and 'WAEC District Officer'. Other occupations associated with small holdings were gamekeeper and rabbit catcher, which might also indicate that they were associated with a local estate. Other farmers were involved in activities such as haulage contractor, building contractor, postmaster or innkeeper. There were a number of professions given, such as doctor, two MPs, publisher of Penguin Books, and an accountant. Various people were indicated as retired, including a retired Colonel. In the *Sussex Sample*, as opposed to the *National Sample*, there appear to be more people in the professions or who were landowners but who also farmed a holding (they were not necessarily any less full time). One particular instance gave 'Foreign Secretary' as the other occupation of Anthony

Table 5.4. Classification of holders as to full-time, part-time, etc.

Name of database	All holders in Tenure Table	Full-time	Part-time	Spare-time	Hobby farmer	Other
National Sample	1226	928 (75.7)	137 (11.2)	102 (8.3)	27 (2.2)	32 (2.6)
Sussex Sample	1055	736 (69.8)	132 (12.5)	54 (5.1)	77 (7.3)	56 (5.3)
Midlands Sample	410	274 (66.8)	55 (13.4)	47 (11.5)	8 (2.0)	26 (6.3)

Percentages in parentheses.

Eden, who held a farm of 5 acres in West Dean, West Sussex.[2] In the *Midlands Sample* the other occupations include fewer professions but more associated with local non-agricultural industries, such as 'director plaster company', 'gypsum miner', 'sand quarrying', 'colliery owner', 'collier' and 'coal merchant'. In Nottinghamshire there were a number of holders whose other occupations were given as 'Co-op Society' and 'UD Council'.

Does farmer occupy other land?
This question invited a yes/no response by entering a cross in a box. In the *National Sample*, 255 holders (20.8%) stated that they occupied other land. In the *Sussex Sample*, the figure was 262 (24.8%); in the *Midlands Sample* it was 65 (15.9%).

[Details of other land]
There was considerable confusion amongst surveyors as to what information should be entered here. The form provides space for details about 'Name of Holding', 'County' and 'Parish' for up to five pieces of other land. Some surveyors gave details of the parts that made up a combined holding, while others merely entered information on other land owned that made up entirely separate farming units. In the latter case, sometimes the farm code for such holdings would be added. Sometimes details of acreage were added, and at the foot of the box comments made such as 'Included in above', or 'Included in September Doomsday' – meaning that information on the holdings detailed here is included in the Primary Return. The name of holdings was sometimes given as 'Land' or 'Land at ...' and in many instances the last two figures of the farm reference were written in pencil in the margin.

Has farmer grazing rights over land not occupied by him?
This question invited a yes/no response by entering a cross in a box.

If so, nature of such rights
The type of grazing rights was added here, e.g. 'Grazing Rights on Ashdown Forest'. Sometimes the acreage involved was given, or even (in one Welsh parish) the numbers of livestock involved. In some cases the name of the person who owned the land where the grazing rights were enjoyed was given, and sometimes the details of the arrangement, e.g. '364 days tenancy'. There are a number of different formats in which the different types of grazing rights were given – for example, 'Common grazing on Wheeler End Common'; '70 acs. grasskeeping let by Sir John Dashwood, at Widdington Park, Lane End'; 'Unlimited Hill Rights'; 'Mountain Rights for 500 sheep'; '400 sheep gates – Nether Moor, 200 sheep gates – Ashap Moor'. The words 'In' or 'Out' were often written in the margin, with no explanation of their meaning. From the number of forms with these kind of marginalia next to the spaces for both 'Other Land' and 'Grazing Rights', it is probable that 'In'

means that the land in question has been included in that Primary Record (as part of the acreage, etc.) although it might be a multiple or combined holding, the recording of which varied regionally. 'Out' would appear to mean that it is not included on that Primary Record.

Section B: Conditions of farm

Proportion (%) of area on which soil is Heavy/Medium/Light/Peaty [Bog or Fen]

The appropriate percentage figures were entered in the relevant boxes for the different categories of soil. A later modification to the 'Peaty' category resulted in instructions to CWAECs to differentiate between 'Bog Peat' and 'Fen Peat': these distinctions were added to the form in manuscript. Very often the percentage given was 100% (e.g. '100% Medium') but differences in the soils of the farm may be detailed (e.g. '35% Heavy, 45% Medium, 20% Light'). Occasionally, the surveyor would enter a cross in one of the boxes, rather than give a percentage figure. In such cases, it can only be assumed that 100% was intended.

Is farm conveniently laid out? Yes/Moderately/No

The appropriate category was indicated by a cross. Occasionally this question was left blank.

Proportion (%) of farm which is naturally Good/Fair/Bad

The appropriate percentage figures were entered in the relevant boxes according to the natural condition of the farm: '100% Good' was much rarer than '100% Fair', and usually the surveyors split the percentages between two, or even the three, categories, for example '20% Good, 75% Fair, 5% Bad'.

Situation in regard to road Good/Fair/Bad

The appropriate category was indicated by a cross.

Situation in regard to railway Good/Fair/Bad

The appropriate category was indicated by a cross.

Condition of farmhouse Good/Fair/Bad

The appropriate category was indicated by a cross. If the holding did not have a farmhouse, this question was often crossed through. Sometimes 'Farmhouse' was crossed out and 'Private House' substituted.

Condition of buildings Good/Fair/Bad

The appropriate category was indicated by a cross. Quite frequently the question was crossed through, in cases where the holding did not possess buildings. Sometimes the words 'Portable' or 'Temporary' were added to 'Buildings'.

CONDITION OF FARM ROADS GOOD/FAIR/BAD
The appropriate category was indicated by a cross. Quite frequently the question was crossed through if the holding did not have any internal roads.

CONDITION OF FENCES GOOD/FAIR/BAD
The appropriate category was indicated by a cross. Quite frequently the question was crossed through if the holding did not have fences. Sometimes the question was answered by two crosses with the word 'to' between them, e.g. 'Good to Fair'.

CONDITION OF DITCHES GOOD/FAIR/BAD
The appropriate category was indicated by a cross. Occasionally the question was crossed through if the holding did not have ditches. Sometimes the question was answered by two crosses with the word 'to' between them, e.g. 'Fair to Bad'.

GENERAL CONDITION OF FIELD DRAINAGE GOOD/FAIR/BAD
The appropriate category was indicated by a cross. Very occasionally the question was crossed through if it was not applicable to the holding.

CONDITION OF COTTAGES GOOD/FAIR/BAD
The appropriate category was indicated by a cross. If the farm did not have attached cottages, the question was crossed through. Sometimes, the number of cottages that were 'Good' or 'Fair' or 'Bad' was entered.

NUMBER OF COTTAGES WITHIN FARM AREA
The appropriate number was entered in the box. Sometimes this box and the next two were left blank, had a short line scored through them or had zero entered to indicate that there were no cottages on the holding.

NUMBER OF COTTAGES ELSEWHERE
The appropriate number was entered in the box. It was rare to get an entry here.

NUMBER OF COTTAGES LET ON SERVICE TENANCY
The appropriate number was entered in the box.

IS THERE INFESTATION WITH RABBITS AND MOLES? YES/NO
A yes or no response was indicated by a cross. Sometimes 'Rabbits' or 'Moles' was underlined, or one of the animals was crossed out, to indicate the particular pest. Sometimes other animals were written in, e.g. 'Foxes' or 'Deer'. Under this and the other headings for the different kinds of pests where there are two (sometimes unrelated) pests, e.g. 'Rooks' and 'Wood Pigeons', sometimes one was crossed out to indicate that it was not part of

the response the surveyor is making. Equally, one may be underlined to indicate that it was the pest the surveyor is returning.

IS THERE INFESTATION WITH RATS AND MICE? YES/NO
A yes or no response was indicated by means of a cross.

IS THERE INFESTATION WITH ROOKS AND WOOD PIGEONS? YES/NO
A yes or no response was indicated by means of a cross. Sometimes 'Rooks' or 'Wood Pigeons' were underlined to indicate the particular bird that was the pest.

IS THERE INFESTATION WITH OTHER BIRDS? YES/NO
A yes or no response was indicated by means of a cross. Sometimes the type of 'Other Bird' was written in, e.g. 'Starlings' or 'Sparrows'.

IS THERE INFESTATION WITH INSECT PESTS? YES/NO
A yes or no response was indicated by means of a cross. Often the type of insect pest was written in, e.g. 'Wireworm'.

IS THERE HEAVY INFESTATION WITH WEEDS? YES/NO
A yes or no response was indicated by means of a cross.

IF SO, KINDS OF WEEDS
The names of the weeds were entered here. Often this was extraordinarily detailed, with very many types of weeds listed. Sometimes further details were given – for example, 'on pasture', 'on arable', 'in one field' – or the relevant OS parcel number.

ARE THERE DERELICT FIELDS? YES/NO
A yes or no response was indicated by means of a cross.

IF SO, ACREAGE
The acreage figure of the derelict fields was entered here. Sometimes the relevant OS parcel numbers were also given.

Section C: Water and electricity

WATER SUPPLY TO FARMHOUSE: PIPE/WELL/ROOF/STREAM/NONE
The appropriate categories were indicated by crosses (one category may be indicated, or several). Sometimes 'Farmhouse' was crossed through and 'Private House' substituted. If the holding did not have a farmhouse, the whole was crossed through. Often other categories of water supply were written in, e.g. 'Pond', 'Spring', 'Reservoir', 'Dyke', 'Ram'. Sometimes further detail was given – for example, 'Pipe – mains', 'gravity from ponds', or 'pumped by hand'.

WATER SUPPLY TO FARM BUILDINGS: PIPE/WELL/ROOF/STREAM/NONE
The appropriate categories were indicated by crosses. (See comments under 'Water supply to farmhouse' above.) Sometimes the words 'Temporary' or 'Portable' were added to 'Buildings'.

WATER SUPPLY TO FIELDS: PIPE/WELL/ROOF/STREAM/NONE
The appropriate categories of water supply were indicated by crosses. (See comments under 'Water supply to farmhouse' above.) Other information – such as 'Water carted to fields', 'Only one field' – was sometimes given. One of the sources may be indicated with a note showing how many fields it went to and sometimes their OS numbers. The additional category of 'Ponds' was permitted by MAF after the Survey was under way. This was often marked by the makeshift construction of another column in the table by hand and then marked with a cross if relevant. However, this was not done consistently, with 'Ponds' sometimes being entered independently. Any of the boxes could be left blank and on a number of occasions there were in fact contradictory responses, where 'None' appears with one or more of the other responses.

IS THERE A SEASONAL SHORTAGE OF WATER? YES/NO
A yes or no response was indicated by a cross. If a seasonal water shortage was indicated for one farm, there was no indication from the records that the problem extended beyond that farm to other adjacent holdings. Sometimes a seasonal shortage was indicated as being relevant to only one or a number of fields. There were also some notes indicating that ponds sometimes go dry.

ELECTRICITY SUPPLY
A yes or no response was indicated by means of a cross for the following questions concerning electricity supply:

- Public light? Yes/No
- Public power? Yes/No
- Private light? Yes/No
- Private power? Yes/No
- Is it used for household purposes? Yes/No
- Is it used for farm purposes? Yes/No.

Section D: Management

IS FARM CLASSIFIED AS A, B OR C?
The appropriate classification was entered here. Sometimes B+ and B– classifications were also used and there were a few cases of C– classifications. The great majority of holdings were given a grading: in the *National Sample* 10 holdings (0.8%) had no grading; in the *Sussex Sample* all holdings were given a grading; while in the *Midlands Sample* one holding (0.3%) was not given a grading.

Reasons for B or C: Old Age/Lack of Capital/Personal Failings

If the classification was B or C, one or more of these categories had to be indicated by a cross. It is not unknown to have all three applying. Sometimes there was no entry under 'Personal Failings' but details were given in the space below. On some occasions when a holding was graded B or C no reason was given for that grading. In the *National Sample* 59 (10.5%) holdings had no reason given. The equivalent figures for the *Sussex Sample* were 12 (1.9%) and for the *Midlands Sample* 5 (1.9%).

If personal failings, details

Here the surveyor entered the nature of the 'personal failings'. Some set terms were used – for example, 'Lack of ambition' or 'No ambition', or 'Lack of initiative', 'Lack of farming knowledge', or 'Divided interests'. Personal comments were rarely included, but these do exist – for example, 'Lazy' or 'Hopeless case'. Also included under 'Personal Failings' were general comments on the farmer that were definitely not failings – for example, 'Gassed in the last war', 'Recently widowed', or 'Failure to cope with impossible conditions'. 'Personal failings' could also be bad farming practices – for example 'Ricks not thatched', 'Insufficient use of fertilizers', 'Too rough and ready: no farming method', or 'Neglect of pasture'. Realities of farming that could not reasonably be described as 'failings' were also entered here – for example, 'Short of labour' or 'Farmer is hampered by the natural poverty of his land'. Many of these comments should probably have been placed the 'General Comments' section on the reverse of the form; in some cases 'See Section E' has been entered in the 'Personal Failings' box.

Condition of arable land: Good/Fair/Poor/Bad

The appropriate category was indicated by a cross (it is important to remember that it is the managerial control over the arable land and pasture, rather than its natural quality, that is being indicated – the conditions marked should correlate with the managerial classification of the farm above). Sometimes more than one category was indicated and the word 'to' added, e.g. 'Fair to Poor'. In other cases the condition was shown by means of percentages, e.g. '40% Good, 60% Poor'.

Condition of pasture: Good/Fair/Poor/Bad

The appropriate category was indicated by a cross. (See comments under 'Condition of arable land' above.)

Use of fertilisers on arable land: Adequate/To some extent/Not at all

The appropriate category was indicated by a cross. Sometimes comments such as 'Very little' or 'Only F.Y.M.' (farmyard manure) were added. On a number of occasions there was an additional cross in one of the boxes, often to one side, and a note in the margin 'Lime', indicating that lime was being added to the soil to the extent indicated by the box.

USE OF FERTILISERS ON GRASS LAND: ADEQUATE/TO SOME EXTENT/NOT AT ALL
The appropriate category was indicated by a cross. (See comments under
'Use of fertilizers on arable land' above.)

*The following details were returned in the box at the bottom right-hand cor-
ner of the Primary Record, after the Management section.*

FIELD INFORMATION RECORDED BY
The name of the CWAEC surveyor should be entered here. It is seldom that
the name here was the actual signature of the surveyor; usually it will have
been copied or even stamped (the latter noted in Dorset, but not in Sussex).
Sometimes the names of more than one surveyor were given (three have
been noted). While it is possible that all these individuals visited the farm,
it is probable that one (or more) of the names was responsible for checking
the data in the office. All names noted are male.

DATE OF RECORDING
The date entered here should be the date that the field visit to the farm was
made for the purposes of the survey. Sometimes more than one day was
given, e.g. '18–19 January 1942'.

THIS PRIMARY RECORD COMPLETED BY
The signature here should be that of the Provincial Advisory Centre staff
member who copied the record. Sometimes the word 'completed' was
crossed out and 'copied' substituted. However, usually if there is a third
signature added (see below), the name is that of a further CWAEC officer
who checked the record. All names noted, bar one, are male. In Dorset
examples, the name stamped here is the same as that of the surveyor: this
must be because for these records the Primary Return form was used as a
field sheet and the record was not subsequently copied.

DATE
The date here is that of the completion of the Primary Return record. Many
seem to have the same date (presumably a large number of records were
copied and completed each day). The date should always be after that of
the 'Date of Recording', although sometimes an obvious error is made
(usually with the year) that puts it earlier. In many cases this date is within
a few days of the 'Date of Recording', although it can also be as much as a
year later.

[THIRD SIGNATURE]
In the *Sussex Sample*, often a third signature (or initials) was added, with a
date. This signature was that of the Advisory Centre staff member who
copied the record.

Section E: General comments

In this section on the reverse of the form the surveyor could enter further details of the farm and its management. Of the 1238 holdings having a Primary Record in the *National Sample*, just under half (611; 49.4%) returned some kind of entry under General Comments. There was a considerable difference between the other two samples. In the *Sussex Sample* as many as 862 (81.6%) returned some kind of entry under this section, while less than a quarter (103; 24.8%) of the 416 holdings having a Primary Record in the *Midlands Sample* had an entry.

Many of the records in the *Sussex Sample* have immensely detailed entries which form a vivid word picture of the farms of the county. Some entries, however, are much shorter, or the section is left totally blank. A photograph of an example comment is shown in Fig. 5.1. Whole parishes of some counties have no entry made here at all, or at the most have an entry of a few words only, e.g. 'Small dairy farm'. The entries of some surveyors follow a common pattern – for example, numbers of livestock and of tractors, with details of rent and rates, are given at the top, with the description of the farm underneath. Some entries include lists of the livestock, and some provide details of OS parcel numbers and the condition of

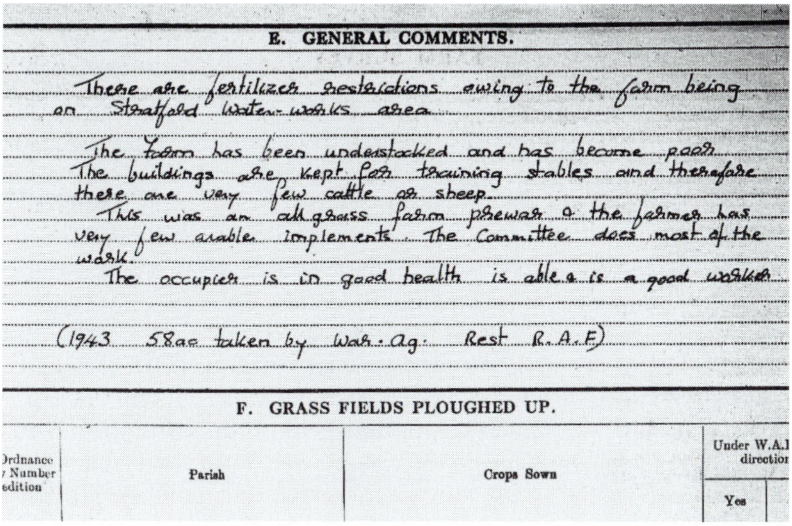
E. GENERAL COMMENTS.

There are fertilizer restrictions owing to the farm being on Stratford Water-works area.

The farm has been understocked and has become poor. The buildings are kept for training stables and therefore there are very few cattle or sheep.

This was an all grass farm prewar & the farmer has very few arable implements. The Committee does most of the work.

The occupier is in good health is able & is a good worker.

(1943 58ac taken by War. Ag. Rest R.A.F)

F. GRASS FIELDS PLOUGHED UP.

)rdnance / Number edition	Parish	Crops Sown	Under W.A.I directior	
			Yes	

Fig. 5.1. Section E of the Primary Record: General Comments. An example of the comments that could be inserted at this point. The comments relate to a holding in Snitterfield, Warwickshire (Source: PRO MAF 32/964 (pt. 1)/240 (273), is Crown copyright and is reproduced with the permission of the Controller of Her Majesty's Stationery Office).

the individual fields. One surveyor invariably started his entries: 'Buildings good, crops good' or 'Buildings good, crops fair'; another was intent in each entry on recording the condition of the farmhouse, and little else. In addition, in East Sussex further information on rent and rates and on numbers of livestock and tractors may be added to the 'Farm Details' section of the Primary Return.

The best and fullest entries describe the condition of the farm, its buildings and land, and provide detailed professional comments on the farming practices being carried out and where these should be improved. Recommendations for fields to be ploughed up are sometimes made here. Comment might also be made on the structure of a combined holding, on difficulties arising from its situation and soil, and on problems such as the shortage of labour. Damage by enemy action, or the requisitioning of land by the military, may also be described.

An example of General Comments at their best is provided by the following for Horse Shoe Farm, East Grinstead:[3]

> 13 Milk Cows, 19 Other Cattle, 1 Horse, 76 Pigs, 2 Tractors. Rent £137. The homestead is 1 mile from the road, and situated in a dale, with the fields encircled on the hills, most of which are very undulating. Most of the soil is a medium loam, with sandstone very near the surface in some fields, often obstructing the plough. In wet weather some parts of the soil [are] apt to silt. This farmer has had misfortune with his mixed herd of dairy cows, and the yield is low at the moment, 1200 galls. last month, but 10 heifers due to calve this autumn will bring production to normal. Many of the fields are very rough grazing, and uneven, due to stone having been quarried years ago. Most of the oak trees have been felled giving the fields an open appearance.

In the *Midlands Sample* there are some descriptive General Comments from Calver, Derbyshire; for example: 'Aspect variable. Millstone Grit. Altitude various 600–1000 ft. Natural growth (trees) – none. Electricity: Public supply available – landlord of house will not let tenants install it'; 'Aspect open. Millstone Grit. Altitude 500–700 ft. Natural growth, trees scanty. Poor pastures being improved by re-seeding. Electricity on land – not tapped for farm. Progressive farm. Landlord unprogressive – probably regards farm as building speculation'; 'Aspect W. Millstone Grit. Altitude 800 ft. 4 fields only: 1 on plough, 1 used as builders yard, 2 let to D. Alsop 73/1 for grazing. (OS 126 0.8 acres just ploughed out not under WAEC direction, March 1943). Intended as building investment. Natural growth Nil'. In Nottinghamshire the comments are generally rather sparse, although on a number of occasions the 'General Comments' box is used for annotating other sections of the form, referring to the various sections (e.g. A2, B13). In Herefordshire no particular pattern appeared and the comments were somewhat sparse, albeit with locally interesting ones.

In the *National Sample* there was a very wide variety in the type and amount of material entered in the 'General Comments'. This ranged from nothing at all, sometimes for every holding in a parish, through perfunctory remarks and formulaic comments to the detailed paragraphs shown above. In Cambridgeshire, Isle of Ely and East Suffolk the box is often fully taken up with a list of OS field numbers, acreages and sometimes the owners' names. This may be typed. Although in some parishes the material was entered systematically it is of little use as it was not done everywhere. The CWAEC obviously had the information on file. These three counties all fall within the same Advisory Province (Eastern) and so perhaps it was local policy to include these details. Sometimes the box is used as a space to make annotations relating to other sections.

Section F: Grass fields ploughed up: 1940 harvest/1941 harvest

NB: in a few cases '1940' and '1941', have been altered to '1941' and '1942' respectively (or even '1943'; see Warwickshire, where a form has actually been locally printed over the Primary Record form).

FIELD ORDNANCE SURVEY NUMBER AND EDITION
The OS parcel numbers of the fields ploughed were entered here. Often these were parts of fields, e.g. '115 Pt'. The date of the edition of the relevant OS sheet should be placed above the list of parcel numbers but this is often omitted. Sometimes parcel numbers are bracketed together, pre-sumably if they form part of a single ploughing operation for the same crop. The format of these references varies greatly.

PARISH
The name of the parish in which the ploughed field lies was entered here. Often the acreage of the field was also added here, before the parish name.

CROPS SOWN
The crops grown in the ploughed field were entered here (e.g. 'Wheat', 'Oats' or 'Rape and Mustard'). The phrase 'Approved Crop' was also used. Sometimes not a crop as such but details such as 'Re-seeded' or 'Fallow' were given. Often the acreage of the field was also entered here. In addi-tion, West Sussex records frequently had more information in this column (e.g. 'Certified to M.A.F. 12 acres'). Sometimes this was bracketed to show that the information relates to several fields (parcel numbers). Some or all of the details might be omitted, and on many occasions a figure might be given for the acreage, either of the field or of the crops which were sown.

UNDER W.A.E.C.'S DIRECTION? YES/NO
Almost invariably the answer to this question was 'Yes' (indicated by a cross in the appropriate box). However, if the entry in 'Crops sown' (see above)

provides information on the crop being 'Certified to M.A.F.', then there was usually no Yes/No entry. Sometimes it was left blank.

The Primary Record – some general comments

The Primary Record form for most counties was always entered in manuscript, but in a few counties some of the records at least were typed. All the Primary Records in Gloucestershire, Leicestershire, Isle of Ely and Rutland were typed; some or a few in Shropshire, Essex, East Suffolk and Huntingdonshire were typed. Sometimes they would be typed in coloured inks – for example, red in Leicestershire and green in Shropshire. Standards of legibility vary widely. Most in East and West Sussex appear to have been copied and are neat and legible, although occasionally there are bad errors in place and personal names. Occasionally there are obvious errors in the transcription of the text for the 'General Comments' section, and sometimes the copier has left gaps where he cannot understand what the surveyor intended. The records do appear, at least in part, to have been checked. Often alterations were made or gaps filled in by another hand, or details queried.

Regional variations are apparent. The distance from London to the remoter regions and the perhaps diminishing control of Whitehall seems to have had an impact on the quality of the records, although there are exceptions, such as Cornwall. Also, the less intensive the agricultural production, the poorer the records seem to be. A few records (examples seen are from Dorset) appear to be the original filled in by the surveyor as a field sheet, and not subsequently re-copied. These are very poorly and inadequately completed, and are scruffy and untidy.

There are also variations in the type of forms and the manner in which they are deposited. They are largely in a consistent order in the folders, and it was noticeable when a Parish did not conform to this order. A proportion of each form was missing or not completed, but not consistently for a holding. On the front of many forms there are various coloured marks, which must relate to the system by which the records were matched and selected for the sample for analysis. Some are marked, 'Urgent. Needed for sample'. Some West Sussex records have 'Primary Record C' in the top left-hand corner, an abbreviation whose meaning is not clear.

The Census Return

It is normally the farmer's own return made direct to the CWAECs that is preserved in MAF 32. Sometimes, however, the Census Return in MAF 32 is the copy made at the CWAEC offices. The original Census Return is in the farmer's own handwriting, and varies greatly in legibility and accuracy of entry.

Sometimes it is corrected in red ink, usually for the total acreage and rough grazings figures, and this is presumably work done at the Advisory Centres, rather than at the CWAEC offices, when the forms were being matched.

On the addressograph side of the Census Return for some Counties there is a stamp giving the date on which it was received by the CWAEC, with either the name or initials of the CWAEC itself or one of its sub-committees. This is found, for example, in Pembrokeshire, Shropshire, Cumberland and the Soke of Peterborough.

Crops and grass

The data entered were numerical and although simple errors of arithmetic are an issue there are only a few points on specific questions that require elaboration:

26 VEGETABLES FOR HUMAN CONSUMPTION
The figure entered here should be the same as the total figure given on the Horticultural Return (see below). Sometimes, it is not, often being left blank on one of the forms. In Nottinghamshire it seemed that whenever the Census Return contained an entry under this heading the Horticultural Return was missing. There is no indication as to where it might have gone.

27 ALL OTHER CROPS
Often a named crop would be written in here (e.g. 'Maize'), with or without a figure for the acreage.

33 TOTAL OF ABOVE ITEMS
Sometimes a correction (usually in red ink) in the addition was made to this figure, and often adjustment made to the figures to which it relates, in particular those for 'Permanent Grass'. Very often the 'Total' figure, and that of 'Rough Grazings', is at variance with the details of acreage given by the addressograph stamp (from the Parish Lists) or on the Primary Return.

The figure for total crops and grass gives an opportunity to measure the internal consistency of the data. We have compared this variable with the sum of the items that precede it on the Census Return (items 1–32), which should be the same. In the *National Sample*, 1054 (83.5%) of 1263 holdings provided exactly the same figures. In terms of percentage differences between the figure given and the sum of the other variables, 1211 (95.9%) were within 10% and 1229 (97.3%) were within 20%. The single largest absolute difference was 134 acres and the mean percentage difference was 3.2%. In the *Sussex Sample*, 954 of 1092 holdings provided exactly the same figures (87.4%). In terms of percentage differences between the figure given and the sum of the other variables, 1022 (93.6%) were within 10% and 1037 (95.0%) were within 20%. The single largest absolute difference was 761 acres and the mean percentage difference was 5.5%. In the *Midlands Sample*, 387 (85.8%) of 451 holdings provided exactly the same figures. In terms of percentage differences between the figure given and the sum of the other variables, 403 (89.4%) were within 10% and 410 (90.9%) were within 20%. The single largest absolute difference was 1587 acres and the mean percentage difference was 3.1%.

34 ROUGH GRAZINGS
The comments above apply.

Labour

There were two sets of question on labour in the 4 June 1941 Census Return – one on the Census Return and the other on the Supplementary Form. Undoubtedly this fact led to some confusion in the minds of farmers. Often words like 'Son', 'Wife' or 'Self' are written in, or sometimes more information on 'Casual Workers', such as '2 evacuee helpers' or 'Land Girl'.

The labour figures provide another opportunity to test the internal consistency of the data. The field containing the total number of workers on the Census Return was compared with the sum of the preceding seven fields. In the *National Sample* 41 (3.2%) out of 1276 were not the same and 26 were annotated with comments indicating that a total had not been entered. In the *Sussex Sample* 10 (0.1%) out of 1105 were not the same; in another case a total had not been entered, and in two cases there were comments indicating that the relevant part of the forms was missing. In the *Midlands Sample* 13 (2.9%) out of 456 were not the same and four were annotated with comments indicating that a total had not been entered. All the samples demonstrated a high level of consistency in these variables.

The questions from the Supplementary Form (129–132) that asked whether the workers returned under the Census Return (questions 35–42) were family workers were also checked for internal consistency. When the data was being entered onto our database, the responses to these question seemed to be some of the least consistent and most confused. The figures from the Supplementary Form for male and female full-time workers and male and female casual workers were subtracted from the sums of the relevant answers to Census Return questions 35–42 to see what the differences were[4]. If any of the differences were negative then the data are not internally consistent, as there should never be more under Supplementary Form questions 129–132 than Census Return questions 35–42, the former being a proportion of the latter. The results for the *National Sample* were that 110 (8.6%) of the responses to Supplementary Form question 129 gave negative numbers. The equivalent percentage for question 141 was 11.1%; for question 130, 4.2%; and for question 131, 1.3%. There were similar results for the *Sussex Sample* and *Midlands Sample*. Overall these results confirm that there are inconsistencies between the answers given for labour within the Census Returns. They also show, however, that the great majority of returns were internally consistent.

Livestock

This section was usually completed competently, although there are occasionally uncorrected errors in addition, or the totals are not completed. Occasionally more information was added, e.g. 'Goslings' or 'Barren cows'.

Horses

Occasionally 'Hunters', 'Pony' or 'Ponies' would be written in here.

The Horticultural Return

This was the form used to copy the information from page 2 of the Supplementary Form for the benefit of the CWAECs, who in turn copied it again. The initial copying was done by MAF at St Anne's, apparently by women of the Poultry Section. Generally, for the average arable farm it was data under 'Stocks of Hay and Straw' that was given. However, in many parishes there can be considerable runs of the Horticultural Return with no data at all.

The form is most extensively used for market gardens and other horticultural holdings. The 'Total' figure for 'Small Fruit, Vegetables and Flowers' given under Horticultural Return question 115 should be the same as the figure given under Census Return question 26. Various fields that are cross-referenced in the forms could be compared to measure these differences (Table 5.5).

In the *National Sample*, when Census Return question 26 was compared with Horticultural Return question 115, 1169 holdings (94.8%) had the same figures. When the sum of Horticultural Return questions 81–114 was compared with Census Return question 26, 1157 (93.8%) holdings had the same figure. There were only 18 (1.5%) holdings where the sum of Horticultural Return questions 81–114 did not match the total given under Horticultural Return question 115. If the sum of Horticultural Return questions 81–86, concerning small fruit, was compared with the total given for small fruit under question 87, then all 1233 holdings had the same figure. If the sum of Horticultural Return questions 81–86 was compared with the sum of Census

Table 5.5. Some measures of the internal consistency of the various figures given for acreages of horticultural produce in the National Farm Survey (NFS) records.

Name of database	National Sample	Sussex Sample	Midlands Sample
Number of holdings with C47/SSY, p. 1	1402	1161	454
Number of holdings with C51/SSY, p. 2	1399	1154	445
Number of holdings in queries that provided the percentage figures below	1233	1072	436
Percentage of occasions when the same figure was returned under C47/SSY, p. 1, question 26 and C51/SSY, p. 2, question 115	94.8	90.9	98.6
Percentage of occasions when C47/SSY, p. 1, question 26 was the same as the sum of C51/SSY, p. 2, questions 81–114	93.8	88.6	97.9
Percentage of occasions when C51/SSY, p. 2, question 115 was the same as the sum of C51/SSY, p. 2, questions 81–114	98.5	95.6	98.9
Percentage of occasions when the sum of C51/SSY, p. 2, questions 81–86 (Small Fruit) was the same as C51/SSY, p. 2, question 87 (Total Small Fruit)	100.0	99.6	100.0
Percentage of occasions when the sum of C51/SSY, p. 2, questions 81–86 (Small Fruit) was the same as the sum of C47/SSY, p. 1, questions 24–25 (Orchards with Small Fruit)	99.0	97.7	99.3

Return questions 24 and 25 (which are referred to under Horticultural Return question 87) then 1221 (99%) holdings gave the same figure.

In the *Sussex Sample*, when Census Return question 26 was compared with Horticultural Return question 115, 974 holdings (90.9%) had the same figures. When the sum of Horticultural Return questions 81–114 was compared with Census Return question 26, 950 holdings (88.6%) had the same figure. There were 47 holdings (4.4%) where the sum of Horticultural Return questions 81–114 did not match the total given under Horticultural Return question 115. If the sum of Horticultural Return questions 81–86, concerning small fruit, was compared with the total given for small fruit under question 87, then 1068 holdings (99.6%) had the same figure. If the sum of Horticultural Return questions 81–86 was compared with the sum of Census Return questions 24 and 25 (which are referred to under Horticultural Return question 87) then 1047 holdings (97.7%) gave the same figure. The total number of holdings returning results was 1072.

In the *Midlands Sample*, when Census Return question 26 was compared with Horticultural Return question 115, 430 holdings (98.6%) had the same figures. When the sum of Horticultural Return questions 81–114 was compared with Census Return question 26, 427 holdings (97.9%) had the same figure. There were 5 holdings (1.2%) where the sum of Horticultural Return questions 81–114 did not match the total given under Horticultural Return question 115. If the sum of Horticultural Return questions 81–86, concerning small fruit, was compared with the total given for small fruit under question 87, then all 436 holdings had the same figure. If the sum of Horticultural Return questions 81–86 was compared with the sum of Census Return questions 24 and 25 (which are referred to under Horticultural Return question 87) then 433 holdings (99.3%) gave the same figure. The total number of holdings returning results was 436.

These results show that the various fields displayed a high degree of internal consistency.

NAME OF FARM OR FARMS/NAME AND ADDRESS OF OCCUPIER
As the Horticultural Return data were copied directly from the Supplementary Form that had been completed by the farmer, the information on name and address was often more accurate than that from the Parish Lists addressograph plates. Names and addresses on the Horticultural Return should always also be compared with the details given on the Primary Return.

PREPARED/CHECKED
The names or initials given in these boxes (and with one exception it is always the 'Prepared box' that is completed: the one for 'Checked' is left blank) were those of the copiers of the data on to the Horticultural Return. The names seem always to be those of women, which ties in with the information that the copying of the page 2 returns was carried out by women from MAF's Poultry Section.

The Supplementary Form

This is the original page completed by the farmer that was detached by MAF's staff at St Anne's to be sent to the Advisory Centres. The 'S.F.' in the top left-hand corner of the form stands for 'Supplementary Form'.

LABOUR ON 4TH JUNE (SUPPLEMENTARY QUESTIONS)
This section of labour questions was often left blank. There was possibly confusion in the farmer's mind because he had already entered labour information on page 1 of the form. There was, however, a general disinclination to answer any of the questions on the Supplementary Form (60,000 reminder forms – see below – had to be sent out). Words such as 'Son', 'Father', etc. are often written in. Sometimes a figure was placed at the head of the section, but is not actually an answer to one of the supplementary questions. This appears to be the labour figure carried forward from page 1 of the form.

MOTIVE POWER ON HOLDING ON 4TH JUNE

Regarding the 'Fixed or Portable Engines' on the farm, the most frequent answers relate to 'Oil or Petrol Engines' and to 'Electric Motors'. Answers were given very infrequently in the other categories, although there have been examples of all. 'Horse Power' details were very often not given, or the HP of one engine only was given when there are two or more on the holding. Sometimes details of the make and use of the engines and motors were given. Engines for pumping and working elevator gear were relatively common, as were electric motors for milking machines. Lawn mowers were frequently entered in this section as well.

Of the tractor entries, 'Wheel Tractors' for fieldwork were the most commonly found. The most popular make was Fordson, but International, John Deere and Allis Chalmers occur relatively frequently as well. There were many cases of cars and lorries that had been converted for tractor work – Austins, Morrises and even a Studebaker. There was often uncertainty about the horse power of the tractor, and Fordsons vary between 14 HP and 35 HP. The HP of the International is frequently stated as 10/20, when this appears to be a model number. There are several instances of the question 'Number in Figures' being answered as the engine number of the tractor, or, in one case, the licence number! Of the 'Track-laying Tractors', Cletrac and Caterpillar were the names most frequently met.

RENT

There was considerable confusion in the completion of these questions. Often the details are entered in another hand, with the word 'Estimate' alongside (this presumably was the work of either the Advisory Centres or the Crop Reporters). Owners of holdings frequently did not know the estimated annual rental value of their land, and would state this fact on the form. Quite often references to 'Schedule A' and 'Schedule B' would be added. Occupiers who were both owners and tenants would sometimes give the names of the holdings or part holdings in question, with their acreages. Quite frequently acreage figures were entered with no corresponding rental value, or vice versa. Sometimes the rent would simply be expressed as, say, '£1 an acre'. There are some cases of quite lengthy annotations in this section of the form from farmers at pains to show how the questions did not apply to their circumstances, or, in one case, to complain about a neighbour!

LENGTH OF OCCUPATION OF THE HOLDING

The years of occupation question was not infrequently found answered as, say, 'All my life', 'Since 1921', etc. There also seems to have been a confusion between the length of time the holding had been in a family's possession rather than that of the present representative of that family. In some instances, where a farmer was not sure what information exactly was required, he has supplied two figures – one, for example, including the period of occupation of his father before him and one stating his own period of management of the farm.

The Parts 1, 2 and 3 entries were sometimes extended in manuscript to Part 4 and even Part 5. Sometimes just an acreage figure was given here and no years, or vice versa. Sometimes an entry would be made for Part 2 while Parts 1 and 3 were left blank.

The 4 June 1941 Census Returns – some general comments

There were no instructions on any of the various forms for the farmer who wants to make a 'nil' or 'not applicable' reply to a question. Most farmers simply left the space blank, but, with the Supplementary Form questions in particular, this led to confusion as to whether the farmer had meant a nil return or whether he had simply omitted to answer the question. Some farmers did, in fact, write in 'none' or 'nil', or scored a line through the box. A few farmers used 'none' and 'one' alongside each other, and care is needed to read such forms. The use of words instead of figures is rare, but some farmers did have a tendency at times to mix the two in sections such as 'Labour'.

Almost all the forms were entered in ink. Occasionally, there are blots on the forms, or a number has run on the poor quality absorbent paper. A very few forms were filled in pencil. The reverse of the Census Return (either the original or the copy made by the CWAECs) was often annotated with comments on the holding or the state of the farm record. 'No PR' and 'PR not asked for' were frequent annotations, as were comments such as 'Taken over' or 'Occupied by Military'.

The Reminder Form (on 'White' Paper) [Form C69/SSY, p. 3: 398/SS]

This was the reminder form sent out to some 60,000 holdings who had either neglected to return the Supplementary Form or whose replies to its questions were inadequate. The forms were sent to Crop Reporters for follow-up action, and were often marked with red crosses against the questions for which a reply was required.

The questions on the form were identical to those on the Supplementary Form (see above). However, the occupier had often signed and dated the C69/SSY form, and completed his address, so this can sometimes serve as a useful check on the farm details given in the Primary Return or on the Parish Lists addressograph plates.

Conclusion

Our comparisons, where possible, between the different forms indicate that the internal consistency of the data varies considerably. Some results that can be directly cross-referenced, such as the acreages for horticultural produce in Census Return question 26 and Horticultural Return question 115, produced high rates of consistency (see Table 5.5). Other variables appear to be less reliable. There was a high level of difference between the Primary Record acreage figure and the sum of the 'Crops and Grass' and 'Rough Grazings' acreages of the Census Return, and there is no indication as to which can be taken as the more reliable (see Table 7.3). The acreage of holdings is fundamental to almost any analysis undertaken using the NFS data, and it is the level of internal consistency of the acreage figures which comprises the most substantial reason for exercising caution in drawing conclusions based on the data.

From these various measures of internal consistency, it would seem that the Census Return and Supplementary Form are generally more consistent with each other than either of them is with the Primary Record, which would be expected from the organization of the Survey as they were checked by the same people and to some extent cross-referenced in that process of checking. There does not seem to have been the same level of effort in checking the Primary Record in relation to them. However, acreages were only given on a minority of the Supplementary Forms, so this also might help to explain the lower rate of difference, as well as making the form less useful in this respect.

Generally, the data from the Supplementary Form seems to be substantially less reliable than that from the other forms. This may well be because it was an innovation within what was a fairly longstanding bureaucratic tradition with which the farmers were familiar – filling in the quarterly farm census forms. The Supplementary Form was new, unfamiliar, poorly designed and poorly worded. It was also asking for additional information from farmers who may well already have come to resent the bureaucratic intrusions of MAF in ways that they did not always see as enhancing production, which was the primary stated aim of wartime agricultural policy.

The quality of the data was a problem recognized by those responsible for its initial analysis even while it was still being collected. The following two quotations from one of the main administrative files of the Survey, show the level of concern:

> for many individual holdings the answering of the remaining questions on page 3, and on pages 1 and 2, leaves much to be desired.[5]

> with regard to the general scope of the national analysis he [*Thomas*] felt strongly that considering the nature of the basic material and the lack of uniformity in the way it was being obtained, any very elaborate analysis would not be justified, and on the whole he thought that it should be largely, if not

entirely, confined to simple summarisation. Further, much more importance should be attached to the factual material in Section A ('Tenure') of the record, than to the more qualitative material such as Section D ('Management').[6]

At a conference of advisory economists held in December 1943 there was discussion about the best way to make use of the vast amount of data collected for the NFS.[7] One economist, Dr Dawe, 'felt strongly' that Advisory Economists should 'use the Farm Survey material as the basis for a comprehensive national report describing the Social Structure of Agriculture'. He argued that the NFS could be used to enhance the professional reputation of agricultural economists who

> were little known outside their own professional circle, and consequently were not sufficiently consulted by the many authorities dealing with the innumerable issues and problems with which the agricultural industry was faced now, and would be faced to an even greater extent after the war. The Farm Survey provided a good opportunity for agricultural economists to make, as it were, their public debut.

Dawe's view was that, unless the agricultural economists took control of the NFS data, 'others less qualified would be consulted about matters which were the economists' own sphere of work [there were indeed signs of this happening already] and they would continue to fail in their obligations to the general public'. The broad consensus of the conference, however, was that much of the NFS data 'provided too shaky a foundation for an agricultural economists' magnum opus'.

One result of this concern was that when the NFS data were sampled to produce the 1946 Summary Report, the sample was substantially biased in favour of large farms for which there were complete and clear records. Our analysis in many ways parallels that of the statisticians and academics who first tried to summarize the data and we have experienced similar difficulties to them in trying to standardize the data and avoid building error into our analysis.

Some general items of concern that have become apparent in our transcription of data are that:

- some farms do not have code numbers;
- the Census Returns for some farms are present without the Primary Record;
- there are regional and local inconsistencies in the completion of the forms;
- it is impossible to tell what corrections were made by the Advisory Economists;
- a proportion of records are missing, ambiguous, illegible or demonstrably incorrect;
- there are several internal inconsistencies within the data, such as the acreage of holdings as found on the Primary Return and the Census Return.

This chapter has described the nature of the data in detail and its regional variations and offered some measure of their internal consistency and rate of error. Although the Research Sub-Committee, in its analysis of the Farm Survey material for the 1946 Summary Report, set some limitations on the accuracy of the data, it was circumspect about the problems it had encountered. We have tried to be more sensitive to the historical nature of the data, but we have also had to deal with its inconsistencies and problems in order to transcribe it into a database. We have tried to avoid treating the NFS merely as a reliable source of empirical data from which unproblematic conclusions can be drawn about the actual nature of agriculture at the time. The NFS tells us as much about the process of state intervention in agriculture and the place of farming within the national and the civil service imaginations as it does about the phenomena that were the object of its analysis.

With all the definitions and redefinitions of the initially very vague categories on the Primary Return and the issuing of instructions to re-standardize data gathering, the data collected perhaps become merely a reflection of all these categories, which were as much for analytical convenience as for representational rigour. None of the errors are declared, and decisions made elsewhere about them being 'satisfactory' are almost all that remain visible of this process of selection and evaluation. The data collected are a reflection of the system of administration used to implement the material and the culture of the civil service at the time. The NFS is very illuminating about the theoretical and methodological ideas that were being developed to deal with large bodies of data and which went on to be of importance in the ensuing development of quantitative methods in a variety of disciplines, including geography.

The Survey provides an extraordinary wealth of data which can be used to answer a wide range of research questions, some of which are explored in Chapter 7. But any researcher has to come to grips with the nature of the data themselves and the degree of certainty which can be placed in conclusions drawn from them. The limitations of the Survey as an historical source are quite substantial, but without it there would simply be no information of this coverage, detail and richness.

Notes

1   PRO MAF 32/NK/052/068 (27).
2   PRO MAF32/XW/303/066 (6).
3   PRO MAF 32 XE/11/22 (25) and MAF 32/1006/22.
4   Supplementary Form question 129 was subtracted from Census Return questions 35, 36 and 37; Supplementary Form question 130 was subtracted from Census Return question 33; Supplementary Form question 131 was subtracted from Census Return questions 39 and 40; Supplementary Form question 132 was subtracted from Census Return question 41.
5   MAF 38/207 (39).
6   MAF 38/207 (49).
7   MAF 38/207. Item 64: Minutes of a conference of Advisory Economists held on 14 December 1943.

The National Farm Survey Maps   6

The Primary Returns and Census Returns that make up the bulk of the National Farm Survey (NFS) are by themselves an enormously valuable source for the study of agriculture and rural life in mid-20th century England and Wales. It is, however, the survival of the associated farm maps that makes the NFS unique. No other national survey is comparable in its provision of detailed agricultural and social information that can be tied down to specific farms. In this chapter we explore the variation in the type and quality of NFS maps that are found in class Ministry of Agriculture and Fisheries (MAF) 73 in the Public Record Office (PRO). This class consists of folders containing either Ordnance Survey (OS) 1:2500 (25-inch) sheets photographically reduced to 1:5000 (12.5-inch), or 1:10,560 (6-inch) sheets. A grid of 16 rectangles is stamped on the inside cover of the folders, with crosses in a rectangle to indicate if any sheets are 'Wanting'. Each sheet was meant to have the OS parcel numbers and acreages transcribed on to it if they were not already present on the base sheet.

Preliminary research had shown that the completion of the maps was highly variable (see Chapter 3; also Foot, 1994; Short and Watkins, 1994). Following an initial examination of a range of maps, in conjunction with the documents in MAF 32, a number of categories were devised to assess the quality and completeness of the maps. The map folder(s) and sheets for each sample parish were examined. This enabled us to geo-reference each holding (see below) and to assess the nature and quality of the maps at a regional and national scale. The main variables collected were: the scale, edition and condition of the map sheets; the type of boundaries used to show farm boundaries; the use of shading and special keys; the system of

© CAB *International* 2000. *The National Farm Survey, 1941–1943*
(B. Short, C. Watkins, W. Foot and P. Kinsman)

farm referencing; the use of OS parcel numbers; the date the maps were drawn; whether the map was a copy or original; the authorship of the map; the use of special stamps; the treatment of non-farming land, including military land; the representation of land used by the County War Agricultural Executive Committees (CWAECs); the treatment of fragmented holdings; marginalia and miscellaneous information. The results of this analysis are presented here for the three main samples.

Number of Maps

For the *National Sample*, 112 map folders were examined. Six of the folders contained no maps, or were not traceable within the PRO, despite being in the MAF 73 catalogue (four from Suffolk, two from Monmouthshire). A further 12 folders had some maps missing from the folders. Sometimes this was a single sheet, but on other occasions it could be up to ten sheets missing out of 16. Nine of the folders with sheets missing were from Wales; three of these were from Monmouthshire and two from Merionethshire. All the folders ordered for the Suffolk sample parish were missing. In addition ten folders had a sheet or sheets marked as 'Wanting'; that is, they had been found to be missing when the folders were assembled for storage in the PRO (see Chapter 3). Most of the information given for the *National Sample* below is therefore based on 106 folders. For the *Sussex Sample*, 22 map folders were examined: only a single folder had any map sheets missing, and this was only one sheet. There were distinct differences in the practice of constructing the maps between East Sussex[1] and West Sussex.[2] For the *Midlands Sample*, 23 map folders were examined: two folders had map sheets missing, although this was probably because they were both on the very edges of their counties and so would not have carried any pertinent information.

Scale

In the *National Sample*, 42 (40%) of the folders contained maps of 1:10,560 while 64 (60%) contained maps of 1:2500 reduced to 1:5000. All the folders in the *Sussex Sample* contained maps of 1:10,560. In the *Midlands Sample*, 19 (83%) of the folders contained maps of 1:10,560 and five (all from Herefordshire) contained maps of 1:2500 reduced to 1:5000.

Edition

The editions of the OS used in the *National Sample* varied between 1903 and 1932. Some were second editions and some had special revisions. In

Essex, a number of the sheets were headed 'Special Emergency Edition of 1938 (Not on general sale to the public)'.[3] In Sussex the editions varied between 1911 and 1932. The folders of maps of Derbyshire were all 1923 or 1924; the folders for Herefordshire were all Second Edition 1903 or 1904 or a Revision of 1937; the maps for Nottinghamshire were of a wider variety of dates, but mostly 1921 and 1922.

Condition of Sheet

The maps were graded according to their physical condition from 1 (very good) to 5 (very poor) and notes were made of their condition. Overall most were in reasonable condition and were legible. In the *National Sample*, 24 (21%) were graded as 1. These tended to be well mounted on cloth backing, perhaps lined along their edges with tape, which was not wearing at all; they were generally clean with only the odd sheet suffering any discolouration or dirtiness. Fifty-four (48%) sheets were graded 2 as they were more generally grubby, probably through being handled, and some had corners cropped off or turned over and small tears at their edges. Some were mounted on brown paper, which was beginning to come off, and others were less well bound along their edges, often with green tape, which was also starting to come off. Twenty-four (21%) of the maps were graded 3. These generally had the same problems as those graded 2, but to a greater extent, and sometimes were badly repaired, mounted or bound. The sheets examined in Essex were somewhat anomalous as there were duplicates of sheets in the folder, some of which were obviously much more used and so were in a much worse condition. There were some nasty tears on them and some kind of correcting fluid had been used. The poorer quality sheets would have been graded 4 and the better ones 2, but as a whole the folder was given a grading of 3.

Only three maps were graded 4. Two of these were from Essex, the other was from Herefordshire. These had more substantial defects, such as a hole in one of the sheets, and were very dirty and torn along their edges. The Herefordshire sheets were multiply folded down along their edges and at their corners, as well as being generally curled (probably due to being stored rolled). The map in worst condition and graded 5 was from Shropshire.[4] This was due to having similar defects to those graded 4 but for a higher proportion of sheets in the folder. Repairs to these sheets had added to their poor condition, as the clear tape used to repair sheets that were torn for some length had now discoloured and stained the sheet, rendering it unreadable.

In the *Sussex Sample*, 12 of the folders were graded as 1 (55%) and the remainder were graded 2 due to minor damage or grubbiness through use. In the *Midlands Sample*, 19 of the folders were graded 2, with sheets slightly grubby from handling and some minor damage. Three of the folders were graded 3, containing sheets which were generally grubbier, curled from being stored rolled at some time, poorly stacked together (hence not fitting well within the folder), more worn and repaired with tape, which was now discoloured; they were all from Herefordshire.

Boundaries

In the *National Sample*, 25 (22%) of the map folders had maps with colour-pencil lines, 21 with colour-paint lines and 26 with some form of crayon. Twelve folders had boundaries marked with some kind of ink, most often black but sometimes a variety of colours to distinguish between holdings. The remaining 25% of the folders had holding boundaries shown in various different ways, such as shading up to the printed field boundaries (no boundary line drawn in as such) or symbols such as crosses and circles to indicate a boundary line.

In the *Sussex Sample*, 14 (64%) of the folders contained maps where the boundaries of the holdings were shown by different coloured crayons. These were the folders for East Sussex. The same colour might be used for several different holdings on the same sheet, so it was not necessarily a means of exclusively distinguishing between different holdings, rather a means of merely distinguishing between holdings that were immediately contiguous. The remaining eight folders (those for West Sussex) had the extent of holdings indicated by a green colour wash and administrative boundaries were highlighted by a yellow line. On some sheets various colours indicated certain kinds of holders; for example, where brown and black crayon lines were indicated as areas being farmed by the East Sussex WAEC.[5] On some other sheets military areas were indicated by a boundary line in red ink or crayon.[6]

The Derbyshire folders from the *Midlands Sample* displayed a number of practices for representing boundaries. There could be lines of various colours in pencil or paint following the OS field boundaries, which may then have been washed in various colours. Sometimes there was only a colour wash that followed the OS field boundaries. In Herefordshire there were colour-pencil outlines for the holding boundaries, and on some occasions the parish boundaries were marked or emphasized in colour. In Nottinghamshire the holding boundaries were always colour-pencil outlines.

Shading

The type of shading fell into two main categories: colour washes and pencil outlines (effectively the boundaries of a holding in colour). These were supplemented by other types that could not be generalized. In the *National Sample*, 17 (16%) of the folders had holdings shaded with a colour wash, 20 folders had pencil-outline shading and 28 folders had some other kind of shading. These consisted of hatching, crayoning, diagonal graphite pencil-shading and colour-ink lines. A large proportion (39%) of the folders had no type of shading at all.

In the *Sussex Sample*, the only type of shading on the maps was different colour crayon boundaries, and the green shading which were not a means of exclusively distinguishing between holdings. In the *Midlands Sample*, four of the Derbyshire folders had some or all of the sheets shaded with a colour wash. There were also areas shaded in colour stripes and with colour boundaries only. Some holdings had a boundary of one colour and shading of another. None of the Herefordshire or Nottinghamshire folders had any shading other than pencil-outline boundaries and some diagonal shading of areas with non-agricultural uses in graphite pencil (see below). One sheet in a Nottinghamshire folder had some graphite pencil notes, in various colours, which may indicate that there had been an intention to shade at some point.

Keys

Some kind of explanatory key was found on maps in a minority (12; 11%) of folders in the *National Sample*. These most often consisted of shaded boxes to correspond with the shading on the map sheets, with the relevant codes or names of holdings or names of the holders (or a combination of all three items). There might also be notes in the key about which holdings were being farmed together. Not all the shading in the boxes was necessarily found on that sheet, and vice versa. Sometimes keys were only for specific items marked on the map sheets, such as built-up areas and holdings under 5 acres, or land planted by the Forestry Commission.[7]

In the *Sussex Sample*, eight (36%) of the folders contained sheets where there was some kind of key. These tended to be in the margin of the sheet and consisted of the holdings by case number (a local code) and the name of the holder, with perhaps the name of the holding as well. In the *Midlands Sample*, 12 (52%) of the folders contained one or more sheets that had some kind of key. In the Derbyshire folders these consisted of ruled boxes with the appropriate colour wash within them and the Farm Reference code, name of holder or holding (or any combination of those three). The standard of presentation was rather variable, some being very neatly ruled and shaded, others being rather scruffy. Not all the holdings on the sheet necessarily appeared in the key, or vice versa. The keys could include items such as the extent of the new Ladybower Reservoir or land which was 'Non-agricultural'. In three of the Herefordshire folders there were annotations that could be a key to the code numbers on the maps and that are not farm reference codes. In two of the Nottinghamshire folders there were sheets with shaded blocks with the county code and perhaps the name of the holder.

Farm Reference Type

In the *National Sample*, 43 (41%) of the folders had some variant of the farm reference code as it is found on the addressograph side of the Census Return and the Supplementary Form. Approximately 35% had only the last two figures of the code, or some other abbreviated form. On very few occasions the full farm reference code was given, with the name of the holder(s), the name of the holding and its acreage, all in small neat black ink inside a neatly ruled box, such as in Essex,[8] but this was unusual. Other variations consisted of the county code (or part of it), the name of the farm and/or the name of the holder. Others used a local code system that incorporated the Crop Reporter's number or the CWAEC District number, perhaps alongside another form of code. There is little uniformity nationally, although there do seem to be local (that is, county-based) practices for codes.

In all 14 of the folders for East Sussex in the *Sussex Sample*, part or all of the farm reference code is given on the maps. In the other folders (that is, those for West Sussex) something called the case number is used. These are also the occurrences where there is a key in the margin. In these cases sub-sections of holdings may also be identified by the addition of a lower case letter (such as a, b or c) to the end of the code. In the *Midlands Sample*, the Derbyshire folders all contained maps that had the last two figures of the farm reference code as their method of reference. Some also had the surname and initial of the holder. The Herefordshire folders all contained maps that had the last two figures of the county code (which was also present in the documents of MAF 32) given in pencil of the same colour; in Nottinghamshire the figures were given in graphite pencil.

Ordnance Survey Parcel Numbers

The CWAECs had been instructed to transcribe the OS parcel numbers and acreages on to the sheets if they were using the 1:10,560 editions. Of the 42 folders in the *National Sample* containing 1:10,560 sheets, 41 (98%) had the OS parcel numbers, but only just over three-quarters (76%) had the acreages as well. In one case the parcel numbers were in red ink and the acreage in purple ink.[9] On another the parcel numbers and acreages were photographically reproduced on the 1:10,560 sheet, having been handwritten on to the sheet from which the copy was made.[10] In the *Sussex Sample*, all of the folders contained sheets where the OS parcel numbers and acreages were given, transcribed on to the base sheets by hand, often to a very high standard (Fig. 6.1). Ten (47%) of the folders in the *Midlands Sample* contained maps that had only the OS parcel numbers. The acreages might be missing from only one or a number of the sheets in a folder, however. The remaining 13 (57%) had both OS parcel numbers and acreages. They were printed on the base sheet in the four Herefordshire folders and transcribed by hand on to the remaining 19.

Dating

In the *National Sample*, 41 (37%) of the folders contained at least some maps that were dated. Of these, 24 do not have a full date and so could not be entered properly into the database. The manner in which they were

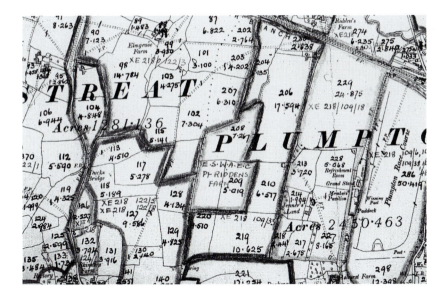

Fig. 6.1. National Farm Survey (NFS) map at a scale of 1:10,560 (Extract). A plethora of numbers on the borders of the parishes of Streat and Plumpton (East Sussex). As well as the acreage figures for the two parishes shown on the base map, the extract also incorporates transcribed parcel Ordnance Survey (OS) numbers, and areas to three decimal points, and individual farm reference numbers (e.g. XE/122/3 Elmgrove Farm), and demonstrates the portrayal of fragmentation for Riddens Farm, part of which is being used under East Sussex War Agricultural Executive Committee (ESWAEC) direction (source: PRO MAF 73/41/39, is Crown copyright and is reproduced with the permission of the Controller of Her Majesty's Stationery Office).

dated varied greatly, and the full dates given ranged from 4 June 1941 to 12 December 1944. These are not necessarily dates of completion, and for a number it is unclear what the date in fact signifies.

Some sheets are dated with only the month, the day and month, the month and year, or only the year. Therefore, the earliest date appearing is February 1940, and the latest date is August 1945. In some folders only a few sheets are dated; in others all the sheets are dated in a uniform manner as a result of a local policy, such as in Cardiganshire, where sheets are dated 'Surveyed Month/Year'. Some sheets have more than one date, such as in Cheshire, where the date when the farm boundaries were surveyed is given, followed by a date when the map sheet was completed.[11]

Some individual holdings are dated on particular sheets, as in Westmorland, where all the sheets are fully dated in the bottom right-hand corner.[12] In the West Riding of Yorkshire the sheets are stamped with the date with a CWAEC stamp.[13] Some sheets are dated in a rather elliptical manner, referring to the dating of the documents in MAF 32. An example is Lincolnshire (Lindsey) where the sheets are marked: 'This map was completed at the same date as the survey forms.' Of the 22 relevant holdings, only one has a date of Primary Return completion (28 September 1942); 19 have a date of field recording of 18 December 1941; one has no date, one has 7 May 1942 and one has 1 November 1942.[14] The dating on some sheets was either obscured by modifications such as binding the edges with linen tape, or completely removed by the cropping of edges or corners of the sheets. None of the folders in the *Sussex Sample* contained sheets that were dated. In the *Midlands Sample*, 12 (52%) of the folders contained maps that were dated in some way. The dates ranged from October 1942 in Herefordshire to August 1944 in Derbyshire. The dates in Derbyshire were accompanied by 'Completed [*date*]'. None of the Nottinghamshire maps were dated. The dates mostly consisted of month and year, and many were obscured wholly or partly by cropping of the sheets or the edges being bound with tape.

Copies

Of the 112 map folders in the *National Sample*, 7 (6%) contained sheets that appeared to be copies rather than the original annotated sheets. Some were photographically reproduced copies of the base 1:10,560 sheets, such as in Northumberland.[15] Others were copies of the NFS sheets with the farm boundaries and details drawn in, such as in Montgomeryshire.[16] They were very difficult to use in terms of locating holdings because all the details were in black and white. It has subsequently been discovered that there were many working copies of these kind of maps, held by District and Technical Officers, which continued to be used in advisory work after the war and which some retired staff still hold.[17] None of the folders in either the *Sussex Sample* or *Midlands Sample* contained sheets that appeared to be copies.

Authorship

Only nine (8%) of the 112 map folders in the *National Sample* contained sheets that had some indication of the authorship of the individual(s) who mapped the Farm Survey information. This largely consisted of initialling and was confined to a few counties: Lincolnshire (Lindsey) and Radnorshire.[18] In Staffordshire one sheet had 'Miss Manby' written in pencil on the reverse, and sheet 11 had 'Mr Meyer'.[19] None of the folders in the *Sussex Sample* contained sheets that gave any indication of authorship. Four of the folders from Derbyshire gave some indication of the authorship of the maps: these consisted only of initials or signatures in pencil which in three of the cases were not properly legible.

Stamps

Over a third (40; 36%) of the folders in the *National Sample* contained sheets that had some sort of stamp on their front or back. Of these, 21 had the PRO piece number stamped on the back, and for 13 of these it was their only stamp. On a number of sheets the ink from the PRO piece number stamp had penetrated to the front of the sheet. Nine folders had a stamp of the firm that had reproduced the base 1:2500 sheets: 'Entwistle Thorpe & Co., London'.

Some stamps on the front of the sheets were the WAEC stamp, sometimes with the address and even the telephone number, such as in Rutland.[20] WAEC stamps would also appear on the rear of some sheets, such as in the East Riding of Yorkshire, where there was also a stamp reading 'Ministry of Agriculture and Fisheries, Recd. 9 Nov 1951. Provincial Office, Leeds'.[21] In Middlesex there was a stamp reading 'Farm Survey 1941–42' with a WAEC stamp. On some sheets there was the address of the Provincial Centre, such as in Northumberland,[22] where 'King's College, Agricultural Economics, Newcastle-upon-Tyne' is given, and in Nottinghamshire, which had 'Midland Agricultural College', both on the back.[23]

In the *Sussex Sample* eight (36%) of the folders contained sheets that were stamped 'Executive Officer, West Sussex WAEC, County Hall, Chichester'. In addition one was stamped 'The Director of Agriculture, West Sussex County Council, County Hall, Chichester'. Rather more (16; 70%) of the folders from the *Midlands Sample* contained maps that had some form of stamp on them. Six had the PRO piece number stamped in the back of some or all of the sheets; these were from Derbyshire or Nottinghamshire.

Three of the folders from Herefordshire had the name of the printer that produced the 1:2500 photographically reduced copies: 'Entwistle Thorpe and Co., London'. All of the folders from Nottinghamshire had 'Midland Agricultural College' stamped on the back; one of these also had 'Notts Education Committee AGRIC' on the front. Six of the Nottinghamshire sheets also had the PRO piece number.

Treatment of Non-farming Land

In addition to boundaries and information specifically about farms, our analysis of the maps shows that there is a considerable number of annotations and additions to the maps of a non-agricultural nature. Many of these relate to features that had changed between the date of survey of the base map and the NFS. In the *National Sample* there are non-farming features marked on sheets in 48 (43%) of the 112 folders. These mostly consist of features such as gravel pits, allotments, newly built-up areas, sewage works, golf courses, common land, smallholdings, waste tips, playing fields, marshes, industrial sites, woodlands, derelict land, new road and railways, or may simply be marked 'Non-agricultural', 'Not agricultural' or 'N.A.'. There are some features that are particular to the regions in which they are set, such as coal workings and tips, plaster mines and sugar beet factories. Other features of interest are the newly developed 'Heathrow. London Airport' in Middlesex[24] and in the Lake District the new level of Wastwater (called High Water on the map sheet) is indicated by a blue line with a note 'New level of lake (approx)' with arrows, and 'Follows 800′ contour'. In addition a new road to the east of the lake is drawn in as two black lines marked 'New road'.[25]

These kind of features do not seem to be marked in any systematic way and so the maps cannot be taken as a full survey of the land they cover, but can be a complementary source to other maps and data from the period. In the *Sussex Sample*, 14 of the folders contain sheets that give some indication of the uses of non-farming land. The features named or indicated include 'Sports Ground' and 'Royhill Holiday Camp'. Several have the term 'Developed'; one has 'Plantation', another 'Allotments' and another 'Built up'. As many as 18 folders (78%) in the *Midlands Sample* contain sheets that include some indication of the uses of non-agricultural land. Five of these use the term 'Non-agricultural' for certain areas of non-farming land and seven use the term 'Built-up' or 'Built on' for certain shaded areas. A variety of features are indicated: ownership by certain people of non-agricultural land,[26] grazing rights on moorland,[27] industrial uses, golf courses, allotments, cemeteries, new roads and railways, gravel pits, ballast holes, gas works, oilfields, forestry and plantations. Some features are peculiar to certain areas such as the Boots factory in Nottingham,[28] Hereford Tiles Ltd[29] and collieries.[30] For Derbyshire there are extensive notes about the grazing rights associated with the Chatsworth Estates. The outline of Ladybower Reservoir in Derbyshire is drawn in by hand and the land around indicated as owned by 'Derwent Valley Water Board (DVWB)'.[31]

Military Land Use

Despite Ministry instructions that ordered military information not to be recorded on the maps, and the possibility of using phrases such as 'Not used for agriculture' to avoid presenting such information (which may account for some of the entries described under 'Treatment of Non-farming Land' above), 21 map folders (19%) in the *National Sample* contain some sheets that presented details of land use by the military. On the whole the information consists of perfunctory notes as to the branch of the armed forces or the responsible government department that is occupying the land. There are five instances of military information that refer to the boundaries of an 'aerodrome' or land being occupied by 'the Air Ministry';[32] in four cases the annotation 'RAF' is written across areas of land.[33] In one Huntingdonshire example there is a lot of detail, taking up most of Sheet

Fig. 6.2. National Farm Survey (NFS) map at scale of 1:10,560 (Extract). Part of the military training area of the South Downs is shown south of the village of West Firle (East Sussex), together with smaller military areas nearer the village itself. (Source: PRO MAF 73/41/67, is Crown copyright and is reproduced with the permission of the Controller of Her Majesty's Stationery Office.) Figure 7.8 shows the broader context.

4, marked 'Air Min' and a straight blue pencil line ruled across it from north-west to south-east labelled 'MAIN RUNWAY'. There is also a yellow continuation of the line in the north-western corner labelled 'PROPOSED EXTENSION scheme a), PROPOSED OVERSHOOT scheme b)'. Land nearby is labelled 'Additional land for scheme a)' and 'Additional land/scheme b)'.[34]

Other sheets have areas of land marked 'Army', 'WD' (probably 'War Department'), 'Military Camp' or 'Military'. Two sheets have military features as part of their base print.[35] A sheet in a folder from Huntingdonshire had an area marked 'POW camps'.[36] One sheet has an area marked 'Ministry of Supply'.[37] In the *Sussex Sample*, 13 (60%) of the folders contain sheets that include some reference to military use of land, as well as the presence of the military being indicated by areas bounded in red ink. There are numerous annotations, often again in red ink, such as 'Military', 'Military Area', 'WD Dump', 'WD' or 'RAF' (Fig. 6.2). Only two (9%) of the *Midlands Sample* folders contain sheets that include some reference to military uses of land. They consist of Derbyshire maps where 'RAF' is marked on areas of Ramsley Moor,[38] and two fields in Nottinghamshire that are marked 'Army' near Norwood Park.[39] See Chapter 7 for a fuller discussion of information on the military use of land.

WAEC Land Use

Relatively few of the maps have land used by the CWAECs. In the *National Sample*, 12 (11%) of the map folders contained some sheets having such information. Fields or bounded areas are marked with the WAEC's initials: for example, EWAEC in Essex or WWAEC in Warwickshire.[40] The Essex maps seem to contain more instances of this information than those of other counties examined. In the *Sussex Sample* a rather greater proportion (12; 55%) of the folders contain sheets that include some reference to the use of land by the WAEC. This usually consists of an annotation of ESWAEC (East Sussex), WSWAEC (West Sussex), KWAEC (Kent), SWAEC (Surrey) or WAEC, according to the county. This could indicate all or part of a holding and might have additional details in the margin. One sheet in a folder has two areas marked 'Brighton sub-committee ESWAEC'.[41] Only one folder from the *Midlands Sample* contains any sheets that make any reference to WAEC use of land. In Derbyshire, holding 79/5 has 'DWAEC' as the name of the holder.[42]

Representation of Fragmented Holdings

One problem that faced all the CWAECs was how to represent links between fragmented holdings on the maps. The variety of these types of linkage are summarized below. Local practice varied greatly, as precise instructions were not issued on this aspect of the construction of the maps, but the variety of methods is very much a variation on a few themes. Of course the shading types described above are a means of demonstrating the distribution of a holding and so any indication of linkages should be considered to be a supplement to the shading for the sake of clarity. In the *National Sample*, 61 (54%) of the folders contain maps that in some way represent links between fragmented holdings or indicate that a holding continues on another map sheet.

There is a variety of methods of indicating linkages: curved lines across features that bisect holdings such as roads or railways;[43] arrows and brackets;[44] 'Pt.' used with code where holdings are fragmented;[45] codes in margins with keys indicating where holdings continue on another sheet;[46] reference to other map sheets on holding;[47] arrows from holdings on the edges of the sheets to annotation such as 'To 73 NE';[48] and curling lines linking up the parts of a fragmented holding.[49]

In the *Sussex Sample*, 17 (77%) of the folders contain sheets that in some way represent the links between fragmented holdings. This was done in East Sussex with arrows linking sections of the holding, most often in red ink with arrowheads at either end. In West Sussex there was not so much a particular method of showing linkages; instead subsections of holdings are marked with the code number followed by sequential lower case letters, i.e. a, b, c, etc. Also in West Sussex, there are separate areas of holdings linked together by red ink brackets. In the *Midlands Sample*, 12 (52%) of the folders contain sheets that in some way indicate the position of various parts of a fragmented holding or holdings that extend over more than one map sheet. Various methods were used to indicate the different parts of a fragmented holding. In Derbyshire there are sometimes codes in the margins referring to other sheets when holdings occur on the edges of sheets. There are also curved and straight lines linking up separate parts of holdings that are bisected by a feature such as a road or railway. There are notes saying 'See also [*map reference*]' referring to other parts of a fragmented holding, often with lists of codes in the margins. In Herefordshire, where a holding only slightly intrudes into a sheet there may be an arrow pointing to the area of land with the code or name of the holding. In Nottinghamshire there are codes and arrows around the edges of the sheets to indicate when a holding goes on to another sheet. There are also some notes in red ink when land in Nottinghamshire is held by holders resident in Leicestershire.

Marginalia

Many maps display some form of annotation or marking in the margins of the sheets other than printed marginalia such as dates or scales. These are often numbers of other sheets or dates with no indication as to what they referred. In the *National Sample*, 47 (42%) of the folders contain maps marked with such marginalia. There are also often names of farms or farmers or lists of farm reference codes.[50] There are notes concerning grazing rights in some areas, such as Derbyshire.[51] In some folders all the sheets are headed 'Farm Survey'[52] and others have the edition of the sheet or its number written in by hand.[53] There may also be codes in the margins and arrows from them to the relevant portion of land.[54] There may be the name or initials of the WAEC and some kind of reference code for the map itself, as in Oxfordshire and Northamptonshire,[55] although this is not clear. In Lincolnshire (Lindsey) there are notes in the margins of some of the sheets relating to market gardens and holdings under 5 acres in adjoining parishes.[56] There are often illegible pencil notes, which probably concern the construction of the map itself.

All the *Sussex Sample* folders contained sheets that had some form of marginalia. These consisted of codes for farms on adjoining sheets, details of WAEC land (see above), additional reference codes including the Ministry Farm Reference code, references to other map sheets that are not OS references (perhaps some local reference system), various miscellaneous annotations and WAEC stamps (see above). Again, with the *Midlands Sample* all the folders contain some sheets that have some form of marginalia. These range from lists of codes with arrows to the holdings and a heading of 'FARM SURVEY' on each of the sheets in Derbyshire to annotations such as 'Small lots with a bungalow on each' and details of land ownership and grazing rights. There are also some notes and codes that do not refer to anything that can be discerned from the map, such as 'Query no. 12'. Various features such as 'Lenton Abbey Housing Estate', 'Allotments' and 'new road' that are not indicated on the sheet may also be marked in the margins.

Miscellaneous Information

The majority of folders (100; 89%) in the *National Sample* contain maps that have some feature or information that cannot be adequately explained under the other headings. This information is difficult to summarize, and often consists of important points of interest about the maps, such as county council ownership of a large proportion of land covered by small-holdings in Bedfordshire,[57] the dominance of a few large landowners such as 'Buckworth Farming Co.', 'Upton Estates' and the Ecclesiastical Commissioners in Huntingdonshire,[58] or grazing rights in Derbyshire.[59] One addition to maps from Lancashire shows how they had been used subsequent to the completion of the Survey: 'Cleared for sand extraction on 30.11.50 subject to restoration'.[60] The Lancashire sheets contain a wealth of well presented information over and above the requirements of the Survey. Other well-drawn sheets were from Somerset.[61]

Some items we included in the database concern the condition or methods of preservation of the sheets, such as binding their edges with linen tape (this was very common, an example being Cambridgeshire)[62] or the punching of holes or slots along one edge so that they could be bound in volumes in some way (Derbyshire)[63] or the pasting of sheets on to boards (Northumberland).[64] Others would concern the quality of the drawing on the sheets, their legibility and usability. Some maps were notable for the poverty of their information and their poor construction. In Middlesex the sheets were completed purely in red ink, which made them very difficult to interpret.[65] The maps from Shropshire were nicely shaded but had inconsistent colours and patchy referencing and were in quite a bad state generally, making it very difficult to geo-reference holdings.[66] Sheets from Westmorland were tidy and well preserved, with clear boundaries and distinctly identifiable holdings, but they contained no information other than the boundaries of the holdings.[67] This was the information that was required for the Survey, but it is all the additional information, the clues about the manner of their construction and the place of the maps in the work of the WAECs that make the majority of other maps so interesting.

Sheets from the Soke of Peterborough have notes that they have been compared with sheets surveyed as part of the work of the Kesteven (Lincolnshire) WAEC and that the two sets of maps agree.[68] In the *Sussex Sample*, 16 folders contain sheets for which some additional miscellaneous information is recorded. This consists of a number of features on the maps themselves, such as district numbers, the names of tenants or holders, some indications of woodlands and forestry, changes in the tenure of holdings during the period of the Survey, and the style of presentation of the farm name. In the *Midlands Sample*, all 23 folders contain sheets for which some item of additional miscellaneous information is recorded. In Derbyshire this often concerns the details of grazing rights. There were also some issues concerning the usability of the sheets. Although the Derbyshire sheets were on the whole well constructed and clear, the quality of the information is variable and sometimes keys to the shading of holdings are absent, incomplete or incorrect.[69] The Derbyshire sheets were often bound along their edges with cloth tape and have punched holes along their left-hand margin for storage in some kind of binder. The Herefordshire sheets are difficult to use; the holdings stretch across many sheets and are poorly drawn. The Nottinghamshire sheets were also bound with cloth tape, and contain some interesting features such as the open field system around Laxton[70] and the site of a new Royal Ordnance Factory near Ruddington, which, although not marked on the maps as such, is discernible from the presence of buildings and a rail depot and references to the requisition of land from farmers mentioned in the documents of MAF 32.[71]

Geo-referencing

An attempt was made to provide a geo-reference for holdings in the samples. Such a reference would allow data from the NFS to be compared with data from a wide variety of other sources and also allow analysis through the use of geographical information systems (GIS). Of 1330 records of farms over 5 acres in the *National Sample* we attempted to geo-reference 1130 by taking a centroid in the single largest portion of the farm as it was identified on the maps. We found it was possible to geo-reference 869 (77%) of these holdings. One simple reason why a holding could not be referenced was the lack of a map (8%). The remaining 15% of problems was due to the absence of codes altogether, map sheets that were completely blank, different or incorrect codes in the maps or documents, or simply not being able to locate the holdings.

This exercise was carried out as a trial to see whether the data could profitably be attached to some form of mapping or GIS software. As some of the grid references were derived from holdings where the codes were not fully present, or details such as name of holding or holder were the only information on the sheets with which to locate them, a proportion of these geo-references must be treated with caution. It is likely that only about 70% of the holdings from this sample could be confidently geo-referenced. Of the 1186 records over 5 acres in the *Sussex Sample*, we attempted to geo-reference 1085. Of these holdings it was possible to geo-reference 931 (86%). It is likely that 80% of holdings in this sample could be confidently geo-referenced. In the *Midlands Sample* we attempted to geo-reference all 477 records over 5 acres. It was possible to geo-reference 391 (82%) of them. It is likely that 77% of holdings in this sample could be confidently geo-referenced in the manner described above. The exercise showed that it is possible to provide a geo-reference for over three-quarters of the holdings.

Conclusion

Our analysis of the NFS maps confirms that they form an extraordinarily important repository of information about rural England and Wales in the mid-20th century. Like most important sources, they are flawed. Although coverage is national, our sample suggests that around 10% of the maps are missing and some counties are more likely to have missing maps than others. A few of the maps are badly damaged, but the overall condition of the sheets is good, with most farm boundaries being clearly legible. One of the surprising results of our analysis was the quantity of information, agricultural and non-agricultural, that could be found on the maps. The various annotations and marginal comments certainly add to the value of the maps as a historical source and can provide a unique insight into land use change in the years immediately preceding the carrying out of the NFS. Our analysis showed that it was feasible to provide a geo-reference for over three-quarters of the holdings. The application of GIS is likely to prove an important means of dealing with the huge amount of data in the NFS archive, and also in linking this data to other information sources to allow for the analysis of changes over time and space. Finally, our assessment shows that the farm boundary information on the maps gives an insight into the nature of farm layout, structure and ownership in the mid-20th century which will prove to be a benchmark for much future research.

Notes

1   PRO MAF 73/41.
2   PRO MAF 73/42.
3   PRO MAF 73/13 (63).
4   PRO MAF 73/53 (62).
5   PRO MAF 73/41 (78).
6   PRO MAF 73/41 (28, 52, 53, 54 and 42/52).
7   PRO MAF 73/25 (39 and 46) Lincolnshire.
8   PRO MAF 73/13 (63).
9   PRO MAF 73/63 (24) Radnorshire.
10 PRO MAF 73/3 (46) Buckinghamshire.
11 PRO MAF 73/6 (48 and 49).
12 PRO MAF 73/44 (12, 19 and 20).
13 PRO MAF 73/49 (98).
14 PRO MAF 73/25 (39) and MAF 32/266 (348).
15 PRO MAF 73/31 (15 and 16).
16 PRO MAF 73/61 (32 and 33).
17 Source: interview with Mike Woods, former Technical Officer with Warwickshire WAEC, 18 October 1995.
18 PRO MAF 73/25 (38, 39, 46 and 47) for Lindsey and MAF 73/63 (24) for Radnorshire.
19 PRO MAF 73/37 (29).
20 PRO MAF 73/34 (8).
21 PRO MAF 73/47 (179).
22 PRO MAF 73/31 (15 and 16).
23 PRO MAF 73/32 (42 and 46).
24 PRO MAF 73/27 (19).
25 PRO MAF 73/44 (20).
26 PRO MAF 73/9 (9).
27 PRO MAF 73/9 (7 and 10).
28 PRO MAF 73/32 (42).
29 PRO MAF 73/17 (33).
30 PRO MAF 73/32 (24).
31 PRO MAF 73/9 (7).
32 PRO MAF 73/1 (27); MAF 73/2 (14); MAF 73/3 (46); MAF 73/19 (12); MAF 73/37 (36).
33 PRO MAF 73/14 (50); MAF 73/33 (27); MAF 73/33 (28); MAF 73/43 (38).
34 PRO MAF 73/19 (17).
35 PRO MAF 73/13 (63) and MAF 73/1 (27).
36 PRO MAF 73/19 (13).
37 PRO MAF 73/10 (18).
38 PRO MAF 73/9 (17).
39 PRO MAF 73/32 (29).
40 PRO MAF 73/13 (64) and MAF 73/54 (3).

41 PRO MAF 73/41 (52).
42 PRO MAF 73/9 (16).
43 PRO MAF 73/2 (14 and 20).
44 PRO MAF 73/6 (48 and 49).
45 PRO MAF 73/7 (72).
46 PRO MAF 73/9 (7).
47 PRO MAF 73/10 (12).
48 PRO MAF 73/13 (64, 72 and 73).
49 PRO MAF 73/14 (50, 51 and 58).
50 PRO MAF 73/9 (6).
51 PRO MAF 73/9 (7).
52 PRO MAF 73/9 (9).
53 PRO MAF 73/17 (25).
54 PRO MAF 73/17 (25 and 32).
55 PRO MAF 73/33 (27) and MAF 73/27 (11 and 12).
56 PRO MAF 73/25 (39).
57 PRO MAF 73/1 (23 and 27).
58 PRO MAF 73/19 (13 and 16).
59 PRO MAF 73/9 (6).
60 PRO MAF 73/21 (84).
61 PRO MAF 73/36 (40).
62 PRO MAF 73/4 (53).
63 PRO MAF 73/9 (6, 7 and 9).
64 PRO MAF 73/31 (15 and 16).
65 PRO MAF 73/27 (19).
66 PRO MAF 73/35 (62).
67 PRO MAF 73/44 (12, 19 and 20).
68 PRO MAF 73/30 (3).
69 See for example PRO MAF 73/9 (16).
70 PRO MAF 73/32 (24).
71 PRO MAF 73/32 (46); see also holdings NM/245/161 (3 and 34) and General Comments in
    MAF 32/367 (161).

Themes in the Wartime Rural Society and Economy of England and Wales 7

In this chapter we examine the potential value of the National Farm Survey (NFS) for the study of themes in the agricultural and social history of mid-20th century England and Wales. The themes we have chosen are land ownership, farm size and structure, mechanization, women and agriculture, the military use of land and the plough-up campaign. These have been selected to demonstrate the different ways in which the NFS data can be used to explore contrasting subjects and to show some of its benefits and pitfalls.

Land Ownership

In this section we explore the extent to which the NFS can be used as a means of determining the ownership of agricultural land in England and Wales in the middle of the 20th century. At the outset it should be realized that the NFS was not designed as a register of land ownership. It was the farm holding and the management of farm units that were the objects of interest. The Primary Return was therefore largely identified by the name of the holder, whether tenant or owner-occupier, as opposed to the name of the landowner. Ownership of land is specifically addressed in Section A ('Tenure') of the Primary Return where two boxes could to be marked indicating whether the holder was an owner-occupier or a tenant, or both owned and rented portions of the farm. Following this, the name of the owner (or owners) of the farm was to be given if the details were different from the name of the holder.

© CAB *International* 2000. *The National Farm Survey, 1941–1943*
(B. Short, C. Watkins, W. Foot and P. Kinsman)

Despite the oblique nature of this information, and some difficulties in utilizing it (which are described below), it is of immense value for any study of land ownership in the 20th century. Theoretically it is possible to find the details of the owner of every agricultural holding in England and Wales over 5 acres in size. The NFS offers a unique picture of land ownership in the middle years of the century, which was in many ways a point of transition for both land ownership and the tenure of agricultural land generally.

Details of ownership

Under Section A ('Tenure') of the Primary Return, the response to the question 'Is occupier tenant or owner?' indicates the tenure of the holding. This question, which requires the placing of a cross in the appropriate box, is sometimes answered by a cross in both boxes, indicating that the occupier is both tenant and owner of different parts of the holding. The relevant acreages are often noted. The answer to this question can therefore show whether a holding is held completely by an owner-occupier, completely by a tenant or whether the tenure of the holding is divided between the two (see Chapter 4, and also the section on farm size and structure later in this chapter, for a more detailed assessment of this question). There then follows the question: 'If tenant, name and address of owner'. The name and address of the owner (or of several owners) is written here. The addresses are sometimes in full, or in an abbreviated form (just the name of the town or village for example), or none is given. In the case of some landowners, the name and address of an agent are given. Where there are several owners, the acreage owned by each is often given.

There are some examples of the details of owners being so lengthy that they are carried over to the 'General Comments' box on the rear of the form. In Little Thetford, Isle of Ely, for example, owners' names and addresses together with lists of the Ordnance Survey (OS) field numbers of the fields they owned are given in this box.[1] However, this level of detail is exceptional and generally surveyors recorded only the information specified in Ministry instructions, with variations according to local practice. Ownership can therefore be derived from various linked variables in the Survey. We analysed the samples to assess how these variables could be used to reconstitute patterns of agricultural land ownership.

Discrepancies and errors

There are remarkably few inconsistencies in the ownership information. In the *National Sample* we found only 12 occasions where a holder was described as an owner-occupier and yet the name and address of an owner was also given. This was probably the result of a copying error. Five of these occurrences were holdings in Ardington, Buckinghamshire, where owner-occupation was given as the tenure of the holder, A.T. Loyd, and the owner was given as 'Lockinge Estate Ltd'. In fact, A.T. Loyd was the owner of the Lockinge Estate (Havinden, 1966). In the *Sussex Sample* there were only two discrepancies of this type. There were also two occurrences where there was a single name for the owner-occupier, but two or more names under names of owners. As the surname was the same for the holder and owners on both occasions, this apparent inconsistency can probably be explained by one individual member of a family being responsible for farming the holding while a number of the members of the family were the owners of the land. In the *Midlands Sample* there was only one occurrence of this kind of contradictory response.

The converse type of error was also measured. In the *National Sample* the holder was given as a tenant, or both a tenant and an owner-occupier, but no name or address of owner was given, on only seven occasions. In the *Sussex Sample* there were only 11 such occurrences and in the *Midlands Sample* there were no occurrences of this kind. These results show that there are very few occasions when the name and address of the landowner cannot be determined from the Primary Return.

Multiple ownership

Multiple ownership of a holding can be ascertained by analysing how many names and addresses of owners were given in Section A, 'Tenure'. Up to nine names and addresses could be specified by surveyors but they were not required to specify the acreage of land each individual owned, nor who owned which parts of the holding. These may have been recorded in this section of the Primary Return, under Section E, 'General Comments' or on the maps, but this would merely be the result of local practice and was not done systematically nor for a very high proportion of the holdings. Arlesey in Bedfordshire was one of the few places in the samples where the acreage owned was given after the name of each person specified as an owner.[2]

Most holdings were owned by a single landowner. In the *National Sample*, 1070 (87%) of the 1226 holdings were owned by a single individual or institution, 91 (7%) by two owners and 65 (5%) by three or more owners. In the *Sussex Sample*, 943 (89%) of the 1055 holdings were owned by a single landowner, 68 (7%) were owned by two owners and 44 (4%) by three or more owners. The equivalent figures for the *Midlands Sample* were 328 (80%) of the 410 holdings owned by a single individual or institution, 53 (13%) owned by two owners and 29 (7%) by three or more owners.

Sometimes the name of the landowner is obscured by the use of the name of a land agent or firm of land agents. This could potentially lead to problems of interpretation if the same agent was used by a number of different landowners in the same area. Moreover, sometimes it is difficult to identify whether an agent's name was put on the form. In some cases the nature of the names given, their addresses and their degree of occurrence in a parish or area would lead one to suspect that it was an agent, although this is not explicit. Generally, the use of an agent's name would indicate the presence of a substantial landowner with several holdings.

Land ownership pattern

Our analysis of the NFS information shows that it can be used to investigate a number of specific points about agricultural land ownership. The main ones are: (i) who owned a holding; (ii) the address of the owner; (iii) whether the holding was held by an owner-occupier or a tenant; (iv) whether it was partly owned and partly rented; (v) whether the farmer held and owned land elsewhere; and (vi) whether the holding was owned by one or a number of individuals or institutions.

However, if there was multiple ownership or mixed tenure, it is not possible to determine which parts of the holding were owned by whom, or how much land was owned by the different owners, unless such details were recorded because of exceptional local practice. Even though the proportions of a holding that were owned and rented were meant to be recorded as the answer to question 145 of the Supplementary Form, the acreage figures from here were found to be much less reliable than those given elsewhere, as the form was new to the farmers and consequently was filled in neither very consistently nor correctly in response to the prompts it gave.

The land ownership information can be cross-referenced to other variables within the NFS data. It is possible to examine, for example, what proportion of agricultural land was owned by owner-occupiers, what was the mean rental value of land under different tenures, and the nature of those owners who rented out their land. Care has to be taken with this type of analysis, especially when dealing with the area of land held by owner-occupiers and tenants on mixed tenure holdings (see Chapter 5, pp. 119–120).

The great advantage of the NFS data is that they can be used to map the pattern of land ownership. Land ownership can be mapped at various scales based on the parish and making use of the records of Ministry of Agriculture and Fisheries (MAF) 32 and MAF 73. The main difficulty is how to deal with those holdings where there is multiple ownership or mixed tenure. Such holdings would have to be indicated in some way such as 'Partly owned by X, Y and Z'. The Survey as a whole is an interesting mixture of precision (such as in most of the Census Return) and ambiguity (such as the 'Tenure' section of the Primary Return), with the result that matching certain items of information together, often in an effort to map them, can be frustrating.

The NFS can certainly be used to map the extent of traditional large landed estates and to examine the type and quality of tenanted farms. Where other sources such as sales brochures and estate and farm records survive, the NFS can be used as corroborative evidence. Indeed the NFS could be used to provide a picture of the variation amongst estates according to their size and location, and indicate the tensions between an increasingly technocratic state, which was beginning to regulate agriculture more closely and was even interested in the idea of nationalizing agricultural land, and some of the most traditionally powerful landholders and farmers on the British agricultural scene.

How can the records be used to reconstruct agricultural estates? It is the names of owners that provide a key to local patterns of land ownership. Often owners of large estates were titled. If a scholar knows the family name of the landowner, possibly linked to the existence of parkland, a known estate name or a home farm, the extent and nature of an estate can usually be readily discerned from the records. The presence of land agents may mask this aspect of ownership, as larger titled landowners would often be represented in the records by the name of an agent or firm of solicitors, but they can sometimes reasonably be assumed to represent a particular owner in an area.

For example, land let out to local farmers on two Herefordshire estates, Foxley and Moccas, could be differentiated according to the name of the owner given. Owners often retained part of the parkland or a home farm. In Mansel Lacy and Yazor most of land was owned by Capt. R.T. Hinckes, who is named as the owner, but also as the farmer of two holdings. One of these is in Mansel Lacy, the name of which was given as 'Accommodation land' and the other in Yazor given as 'Yazor Court'.[3] On the maps in MAF 73, the latter contains the house and surrounds of what is known as the Foxley estate. In contrast, in the parish of Moccas, the name of an agent, H.K. Foster Esq., is present under the name of the owner for most of the rented land in the parish, except one holding called 'Lower Moccas'[4] where the name of the owner is given as Sir Geoffrey Cornewall, who is also named as the holder of Moccas Court.[5] The spelling of many of the above items (e.g. Hin(c)k(e)s and Corn(e)wall), plus the address of the agent H.K. Foster Esq., given as both 29 and 129 St Owen Street, Hereford, is highly variable, although it can be safely assumed that they are all referring to the same items.

Types of tenure and land ownership

The first information on land ownership from the Primary Return, under Section A ('Tenure'), is whether the holder is also the owner. The results are shown in Table 7.1. For our three samples, owner-occupiers made up

Table 7.1. Percentage of number of holdings under different types of tenure.

Name of database	Owner-occupiers (%)	Tenants (%)	Mixed tenure (%)
National Sample	27.7	62.2	10.0
Sussex Sample	33.3	58.4	8.3
Midlands Sample	22.7	65.9	11.5
England and Wales[a]			
(Summary Report 1946)	31.4	62.1	6.5

[a]Source: *National Farm Survey of England and Wales. A Summary Report*, HMSO, London, p. 20 (note that the definitions for the three categories are slightly different).

between 23% and 33% of all holders. These results compare well with the roughly equivalent figure of 32% from the sample used in the Summary Report (1946). If the holder was an owner-occupier, then it is likely that they were the owner and occupier of that single holding only. Owner-occupiers rarely held more than one named farm, although they might also have owned a small amount of land nearby, indicated in Section A under the question, 'Does the farmer occupy other land?' The NFS confirms that the traditional landlord/tenant system continued in the mid-20th century to be the dominant form of agricultural land tenure even after a 20-year period characterized by the sale and breaking up of landed estates. Between 58% and 62% of all holdings were tenanted in our three samples. Holdings of mixed tenure comprised between 8% and 12% of all holders.

Types of landowners

Those landowners discernible in the NFS can be divided into a number of different types and categories. The principal distinction is between private landowners and land owned by institutions.

Private landowners

TRADITIONAL AGRICULTURAL ESTATE
By far the most important kind of land ownership was that of the traditional agricultural estate owned by private individuals, many of whom were titled. These were the most significant single type of landowner, in terms of both the number of holdings and the acreage owned, although a definitive statement concerning this is difficult to give, due to the nature of the acreage figures given for holdings, the lack of data on the proportion of holdings owned by different people when there was multiple ownership or mixed tenure, and the manner in which the names of owners and their social status was indicated.

Some of these traditional estates owned all the land in a parish (as was the case with Beeley, Derbyshire, part of the Chatsworth Estates),[6] or virtually all of it, as did Lord Cawdor at Stackpool Elidor, Pembrokeshire,[7] or Lady Catherine Ashburnham in Ashburnham, East Sussex.[8] A parish might be dominated by one estate holder with a small number of owner-occupiers and individuals who owned a few acres of land, such as in New Radnor, Radnorshire, where Sir H. Duff Gordon was the dominant landowner.[9] Clearly the NFS maps are of great value in reconstructing the land ownership boundaries and can be used to show the extent to which estates were fragmented.

Estates could also be specified in the NFS records by a name for the estate or through the name of a land agent, such as the Shefford Feoffe Estate, Bedfordshire,[10] the Lockinge Estate, Berkshire,[11] and Paget Estates, Nottinghamshire.[12] Chatsworth Estates were concentrated in a few parishes, but extended significantly into others, especially through their upland holdings. Chatsworth Estates were therefore the dominant landowner in a number of the parishes from the Bakewell and Chapel-en-le-Frith districts in our sample including Edale, Beeley, Stoney Middleton and Hope Woodlands,[13] along with the Derwent Valley Water Board, the Forestry Commission and a few private individuals.

Many traditional estates were not very large but were important within a particular parish. For example, Miss M. Edge of Strelley Hall was the single owner of all but one holding in Strelley, Nottinghamshire.[14] Others would be virtually the exclusive landowner in a parish that was the focus of their estate. Capt. R.T. Hinckes, for example, owned most of Mansel Lacy, Herefordshire, and in addition owned a significant acreage in Wormsley and Yazor and had a few other acres in nearby parishes, such as Moccas, Mansel Gamage and Bishopstone.[15]

EXECUTORS

A proportion of land was always in the hands of the executors of wills, at a variety of scales. This was often given as 'The Executors of the late …' or '(The) Exors (of the late) …', sometimes with the land agent, the firm of solicitors and the name of the person they were representing. In the *National Sample* there were 24 occurrences (2%) of the owners of holdings being given as executors. In the *Sussex Sample* there were 22 occurrences (2%) and in the *Midlands Sample* there were 13 occurrences (3%). This is a small but not insignificant proportion of holdings, and within a particular parish or district could have an impact on a high proportion of the holdings if a significant landowner had died, such as in Hamsey, East Sussex, where both Lord Monk Bretton (agents: Strutt & Parker) and Sir Henry Shiffner (also spelt 'Stiffner') (agents: Cobbe & Co.) had recently died and their estates were still in the hands of their executors. Ten of the 13 holdings in that parish were therefore in a suspended state of ownership, and this may well have had some impact upon the management of that land.

SMALLER NON-FARMING PRIVATE LANDOWNER

The NFS records are particularly valuable in examining the importance of the smaller non-farming private landowner. With large estates, it is likely that other records describing the estate may survive. With the smaller landowner, who might own anything from a field, through a few fields to a couple of farms, the likelihood of surviving records is minimal. The NFS allows us to see how many of these landowners existed within particular areas and to examine, for example, their importance in issues such as land development near towns.

Land owned by institutions

LOCAL GOVERNMENT

Local government was a very significant landowner. Many county councils had estates established under the Small Holdings and Allotments Acts of 1907 and 1908 (Wood, 1982). The NFS records provide a snapshot of the extent and form of holdings greater than 5 acres on these estates. Examples from our sample include Arlesey, Bedfordshire,[16] where half of all the holdings in the parish were owned partly or wholly by Bedfordshire County Council. This area was intensively horticultural and 'First Garden City' was given as one of the names of the owners of one holding. Many other parishes in the sample contained lands held by either county councils (for example Shepreth in Cambridgeshire), by the Agriculture Departments of county councils (as in Stanwell, Middlesex), by urban district councils (such as Long Eaton UDC in Stapleford and Beeston, Nottinghamshire) and by parish councils (e.g. Shepreth Parish Council and Foxton Parish Council in Shepreth, Cambridgeshire, and Little Thetford Parish Council, Little Thetford, Isle of Ely).[17]

WATER BOARDS

Water boards could also be significant local landowners, especially where major reservoir construction had taken place, such as in Derbyshire around the Ladybower reservoirs. The Derwent Valley Water Board was one of the major landowners in the parishes of Derwent and Hope Woodlands.[18] In the three parishes of Beeley, Hope Woodlands and Derwent, the NFS records show that the land was divided largely between Chatsworth Estates, the DVWB, the Forestry Commission and one other titled individual. Elsewhere, local water companies held small amounts of land, perhaps next to waste and water processing facilities or adjacent to rivers. An example is a holding owned and let to a farmer by Trent Catchment Board, Stapleford and Beeston, Nottinghamshire.[19]

FORESTRY COMMISSION

The Forestry Commission appears as a landowner in a number of our sample of parishes, particularly in upland areas where afforestation had taken place, or was planned. Woodlands, including new plantations, were excluded from the NFS, but areas of agricultural land yet to be afforested, or laid out as foresters' smallholdings, were included (Watkins, 1984). In Derbyshire several of the Primary Returns that gave the holder as having another occupation that was associated with the Forestry Commission, such as 'woodman', were described as inefficient. Other places where the Forestry Commission was found to be a landowner include Denny Lodge, Hampshire, Machynlleth, Montgomeryshire, and East Grinstead, Sussex.[20]

CROWN ESTATE

The Crown Estate did not often appear in the NFS records we transcribed, but it was a significant landowner in a number of parishes in West and East Sussex. Small areas were owned in Friston and Upper Beeding and the Crown Estate was the dominant landowner in Poynings and Pyecombe.[21] Crown lands may also often have been represented by agents.

NATIONAL TRUST

The National Trust is also distinguished as an owner, although not much of its land appeared in our samples. These included two holdings in Edale, Derbyshire,[22] and a number of holdings in East and West Sussex.

CHURCH OF ENGLAND

In some parishes the Church of England was the single most significant land holder, such as in Leighton, Huntingdonshire, and in others it was one of a number of landowners, such as at Godney, Somerset, Charlton-on-Otmoor, Oxfordshire, and Rhosili, Glamorgan.[23] In England the name of the owner in these cases would be given as 'The Ecclesiastical Commissioners', whereas in Wales it was given as 'The Welsh Church Commissioners'. Other religious institutions also cropped up as minor landowners, such as the Notre Dame Convent in Hartfield, East Sussex, and a Methodist Church in Stapleford and Beeston, Nottinghamshire.[24]

LOCAL CLERGY

Local clergy were often present among a variety of types of minor landowners within parishes. They are normally referred to through the title 'Rev.' or as 'Vicar/Rector of X'. There were 22 occurrences (2%) of this kind in the *National Sample*, 10 (1%) in the *Sussex Sample* and 16 (4%) in the *Midlands Sample*. In the parish of Charlton-on-Otmoor, Oxfordshire, the Rector owned part of four holdings; in the parish of Gotham, Nottinghamshire, the Rev. B.P. Hall owned all or part of five holdings; and in Buckden in the

West Riding of Yorkshire the Rev. Woode and the Rev. Menzies owned five holdings between them.[25]

INDUSTRIAL CONCERNS

The industrial concerns that held agricultural land were normally primary, extractive or agriculture-related and linked to the local occurrence of a natural resource of some kind. Examples include: the British Portland Cement Company in Shepreth, Cambridgeshire, and Arlesey, Bedfordshire; the London Brick Company in Bedfordshire; the Stanton Iron Works Co. in Bilsthorpe, Nottinghamshire; Shipstones Brewery in West Bridgford, Nottinghamshire; and Hansons Brewery in Eastwood, Nottinghamshire. Most of the industrial landowners found were in the *Midlands Sample*. Railway companies also appeared on a few occasions, particularly the LMS Railway in Beeston and Stapleford, Nottinghamshire, and Helpston, Soke of Peterborough, returning small plots of land, often adjacent to the railway lines.[26]

COOPERATIVE SOCIETIES

Cooperative Societies owned land in the *Midlands Sample* where parishes were near or actually part of urbanized areas. In Stapleford and Beeston, Nottinghamshire, the 'Long Eaton Coop Soc' owned and farmed one holding and the 'Nottingham Co-op Society' owned and farmed two holdings – one in Ruddington and one in West Bridgford. They did not own a high proportion of the agricultural land, but stood out as being institutional rather than private owners. Their ownership of land was linked to the retail trade in milk and was used for the keeping of horses to distribute milk bought wholesale locally.[27]

CLUBS AND SOCIETIES

A number of clubs and societies were present as landowners, although on a small scale. The Shepreth Benefit Society, in Shepreth, Cambridgeshire, owned one holding and the Shepreth Club owned another. In Aberystwyth Borough, Cardiganshire, 'The Golf Club' owned a holding, and the Girl Guides Association owned one in East Grinstead, East Sussex.[28]

EDUCATIONAL INSTITUTIONS

Educational institutions owned land occasionally, but could be locally significant. One holding in Stapleford and Beeston was given as partly owned by Church Schools (34 acres). In Helpston, Soke of Peterborough, Christ's College, Cambridge owned all of two holdings and a part of another one, making it an important landowner in that parish.[29]

FINANCIAL INSTITUTIONS AND HOSPITALS

Financial institutions were rare landowners in our samples. We found one holding in the parish of St Just in Roseland, Cornwall, owned by Lloyds

Bank.[30] Hospitals and asylums owned land in a number of parishes. The Three Counties Hospital, in Arlesey, Bedfordshire, a very large psychiatric hospital, owned and farmed a holding, as did Staines Emergency Hospital in Stanwell, Middlesex. In Buxted, East Sussex, one holding was indicated as being owned by 'Trustees of Cottage Hospital'.[31] Trustees are another example of a form of institutional land ownership, but again on a very small scale. They would include trustees for an estate, for individuals or for charities. Sometimes there was an indication of who the trustees were, such as Goreham Trustees and Lloyds Bank.

Conclusion

The NFS clearly provides an enormous amount of information about patterns of land ownership and different types of landowner for the mid-20th century. This information is likely to be most useful for local studies of land ownership, where information from the NFS can be combined with data from other sources. Many large private estates and institutional landowners have such surviving estate records and maps that can be used for corroboration. The NFS is perhaps at its most valuable in providing a complete record of agricultural land ownership in map form. This enables the role of the smaller landowner as a provider of land for farming purposes to be assessed. Overall the NFS information provides an extremely valuable baseline that can be used with the pre-First World War Lloyd George Survey (Short, 1997) and modern surveys to measure changes in land ownership.

Farm Size and Structure

One of the most important themes emerging from the initial analyses of the NFS material was the size of farms in Great Britain and, related to this, the degree of fragmentation into multiple holdings that was to be found.

The sources of farm size and acreage figures

There are a number of opportunities within the NFS returns to record the acreage figures of a holding. Firstly, the total acreage of the holding was supposed to appear on the Primary Return, despite there being no designated space for it. Surveyors were instructed to include it in the top right-hand corner of the Primary Return, above the code number (for which there was also no space designated).[32] In the *National Sample* as many as 1269 (95.4%) of the 1330 holdings collected, do have an acreage figure on the Primary Return. In the *Sussex Sample*, an impressive 1179 (99.4%) of the

1186 holdings collected, have an acreage figure on the Primary Returns. In the *Midlands Sample* 377 (79%) of the 477 holdings collected, have an acreage figure available on the Primary Return. The two main reasons for the acreage figure not appearing in the NFS are: (i) those instances where only the 4 June Return exists and there is no Primary Return extant; and (ii) those occasions that have duplicate Primary Returns from multiple holdings. Remarkably, therefore, although there was no allotted place on the Primary Return, an acreage figure is available for almost all holdings.

Other information on the area of holdings is provided on the Census Returns. On the rear of the Return and the Supplementary Form, as part of the addressograph, are the two figures separated by a slash, which are the figures for total crops and grass and of rough grazings from the previous year's 4 June Returns. On the data side of the C47/SSY (page 1) there are also the total acreages of crops and grass and of rough grazings, under items 33 and 34, respectively. These can provide a total acreage figure for 1941 as well. Moreover, on the data side of the Supplementary Form there is space for the proportions of the acreage of holdings that are both owned and rented. There is also space for the proportions of a holding that have been occupied by the holder for different lengths of time. In those few cases where a holding's total acreage is not available from the Primary Return, therefore, it can be calculated by using the sum of the two figures from the Census Return.

However, to return to the measures of internal consistency as explored in Chapter 5, any derivation of acreage figures must be approached with some caution – as shown by the analysis set out in Tables 7.2 and 7.3.

Of the 1330 holdings collected in the *National Sample*, 1207 (90.8%) have figures for both the Primary Return acreage, and the entry for total crops and grass and rough grazings. If the sum of the latter two is compared with the former, then only 409 (33.9%) are exactly the same; 726 (60.2%) are within 10% of the Primary Return acreage and 857 (71%) within 20% of it. The single largest absolute difference was 1775 acres and the maximum percentage difference was 1355%. The mean percentage difference was actually as much as 25.8%.

This huge difference is mirrored by similar differences in the other two samples. For the *Sussex Sample*, 1086 (91.6%) of the 1186 holdings collected have figures both for the Primary Return acreage and for the total crops and grass and rough grazings. Again, if the sum of the latter two is compared with the former, then only 328 (30.2%) are exactly the same; 608 (56.0%) were within 10% of the Primary Return acreage and 733 (67.5%) within 20% of it. The single largest absolute difference was 1021.2 acres and the maximum percentage difference was 1503.5%. For Sussex, the mean percentage difference was 21.4%. In the *Midlands Sample*, 363 (76.1%) of the 477 holdings collected have figures both for the Primary Return acreage and for the total crops and grass and rough grazings. Once again, if the sum of

Table 7.2. Some measures of the internal consistency of the various figures given for total acreages in the National Farm Survey (NFS) records.

	National Sample	Sussex Sample	Midlands Sample
Primary Returns with acreage figure	95.4	99.4	79.0
Percentage of holdings with both PR acreage and C47/SSY, p. 3, Crops & Grass and Rough Grazings figures	90.8	91.6	76.1
Percentage of holdings where PR acreage and sum of C47/SSY, p. 1, Crops & Grass and Rough Grazings figures were exactly the same	33.9	30.2	42.7
Mean percentage difference between PR acreage and the sum of C47/SSY, p. 1, Crops & Grass and Rough Grazings figures	25.8	21.4	29.3
Percentage of holdings where sum of 398/SS, p. 3, acreages and PR acreage were exactly the same	22.1	22.0	29.0
Mean percentage difference between 398/SS, p. 3, acreages and PR acreage	12.1	21.9	7.3
Maximum percentage difference between 398/SS, p. 3, acreages and PR acreage	189.1	100.0	79.2
Percentage of holdings where sum of 398/SS, p. 3, acreages and sum of C47/SSY, p. 1, Crops & Grass and Rough Grazings were exactly the same	61.0	40.0	61.3
Mean percentage difference between 398/SS, p. 3, acreages and sum of C47/SSY, p. 3, Crops & Grass and Rough Grazings figures	6.8	8.8	12.6
Maximum percentage difference between 398/SS, p. 3, acreages and sum of C47/SSY, p. 1, Crops & Grass and Rough Grazings figures	198.7	100.0	216.7
Percentage of holders returned as Tenants where a figure for Rental Value is returned on 398/SS, p. 3	90.3	89.7	92.0
Percentage of holdings returned as partly rented and partly owned where both acreage figures are given on 398/SS, p. 3	80.1	74.7	80.4

PR = Primary Return.

Table 7.3. Summary table of differences between acreage figures on the Primary Return (PR), Census Return (C47/SSY, p. 1), and Supplementary Form (398/SS, p. 3).

	National Sample	Sussex Sample	Midlands Sample
Percentage of holdings where PR acreage and sum of C47/SSY, p. 1, Crops & Grass and Rough Grazings figures were exactly the same	33.9	30.2	42.7
Mean percentage difference between PR acreage and sum of C47/SSY, p. 1, Crops & Grass and Rough Grazings figures	25.8	21.4	29.3
Percentage of holdings where sum of 398/SS, p. 3, acreages and PR acreage were exactly the same	22.1	22.0	29.0
Mean percentage difference between 398/SS, p. 3, acreages and PR acreage	12.1	21.9	7.3
Percentage of holdings where sum of 398/SS, p. 3, acreages and sum of C47/SSY, p. 1, Crops & Grass and Rough Grazings were exactly the same	61.0	40.0	61.3
Mean percentage difference between 398/SS, p. 3, acreages and sum of C47/SSY, p. 3, Crops & Grass and Rough Grazings figures	6.8	8.8	12.6

the latter two is compared with the former, then only 155 (42.7%) are exactly the same; 235 (64.7%) were within 10% of the Primary Return acreage and 268 (73.8%) within 20% of it. The single largest absolute difference was 465 acres and the maximum percentage difference was 960%. The mean percentage difference for the Midlands was 29.3%.

The acreage figures on the Supplementary Form were much less reliable than those from the standard 4 June Return, as the form was new to the farmers and consequently was completed neither very consistently nor correctly in response to the prompts it gave. In the *National Sample*, 70 holdings (9.7%) that were returned in the 'Tenure' section of the Primary Return as being held by tenants gave no figure for 'Payable Rent'. Of the holdings which were returned as being partly owned and partly rented, 23 (19.2%) gave no figures for acreage or rent. The differences between the acreage figures that could be derived from these variables and the acreages given on the Primary Return and Census Return were then calculated. From the 398/SS, only 78 (63.4%) of the holdings that should have returned acreage figures for both the proportion that was owned and that which was rented did so.

If the acreage derived from these variables is compared with the acreage from the Primary Return, only 17 (22.1%) were exactly the same and 44 (57.1%) were within 10% of the Supplementary Form acreage, and 54 (70.1%) within 20% of it. The single largest absolute difference was 162 acres and the maximum percentage difference was 189.1%. The mean percentage difference was 12.1%. Moreover, if the acreage derived from these fields is compared with the summed acreages for crops and grass and rough grazings from the Census Return, 47 (61%) were exactly the same; 67 (87%) were within 10%. The single largest absolute difference was 620 acres and the maximum percentage difference was 198.72%. The mean percentage difference was 6.8%.

This would seem to indicate that in the *National Sample* the acreage figures derived from the Census Return (C47/SSY, page 1) and Supplementary Form (398/SS, page 3) are more consistent with each other than either of them is with the acreage from the Primary Return. This is not necessarily a measure of their accuracy, however, as they both come from sections of the forms completed by the farmers, who would have been likely to return the same figures. Similar results were obtained from analysis of the data in the *Sussex Sample* and the *Midlands Sample*.

Holding size

Given the caveats above, one must interpret the holding sizes from the NFS with care and an understanding of which particular source is being used. In the 1946 Summary Report there is an analysis of holdings by type of farming and by size, and it does not therefore seem appropriate to repeat such an exercise here. However, Table 7.4 sets out the basic findings of this present investigation on holding size for the National Sample, for Sussex and for the Midlands, and also compares these findings with that same basic information from the 1946 Summary.

The broad correspondence between the size of holdings as derived from the present sample, and those derived from the 1940s 14% sample is strong, although the present analysis is seen to lean more heavily towards the larger farms than that of the contemporary study. Another way of presenting the varying spatial pattern of farm size that was not attempted in the

Table 7.4. Holding size 1941–1943.

Acreage	National Sample		Sussex Sample		Midlands Sample		1946 Report
	No. of holdings	% of total	No. of holdings	% of total	No. of holdings	% of total	% of total
5–25	414	32.7	421	35.7	123	32.6	35.0
25–100	436	34.4	374	31.7	137	36.3	38.0
100–300	324	25.6	241	20.5	97	25.7	22.0
300–700	66	5.2	110	9.3	18	4.8	4.0
700+	27	2.1	32	2.7	2	0.5	1.0
Total	1267		1178		377		

1940s is by Advisory Centre Provinces (Table 7.5). The present sample depicts the smaller holding sizes of the North Western Province compared with the larger holding sizes in Yorkshire – a distribution also depicted, especially for the East Riding, in the 1940s sample (MAF, 1946: 92). However, given that such sizes were derived from one sample parish in each county that made up the Province, the data are not sufficiently robust to take generalizations such as these much further.

Farm fragmentation

Another issue that was analysed at some greater length in the 1940s was that which came from the questions in the survey on the convenience of farm layouts and the degree of fragmentation or 'severence' of holdings. Inspection of the data, once more by Advisory Centre Provinces, demonstrates some spatial variation, and also once again underpins the findings from the 1946 Summary arising from the field inspections of the surveyors (Table 7.6). They were required to give an assessment of the convenience in terms of the shape of the holding, the size, shape and arrangement of its fields, the position of the farmhouse, and the internal arrangement of the farmstead (the distance between the farmhouse and buildings, distance between buildings, convenience of water supply, etc.)

The regional variations in convenience of layout can be seen between the high proportion of positive responses in the East and West Midlands, South East and Yorkshire compared with the lower proportions in the Northern, Western and Welsh Provinces. The national figures once again accord closely with those from the 1940s sample, which were referred to in

Table 7.5. Holding size in Advisory Centre Provinces.

Province	Holding size (acres)				
	5–25	25–100	100–300	300–700	700+
East Midland	37 (32.5)	40 (35.1)	29 (25.4)	7 (6.1)	1 (0.9)
Eastern	46 (28.4)	53 (32.7)	55 (34.0)	7 (4.3)	1 (0.6)
North Western	16 (51.6)	7 (22.6)	8 (25.8)	0	0
Northern	22 (27.2)	33 (40.7)	13 (16.0)	8 (9.9)	5 (6.2)
South Eastern	42 (32.1)	47 (35.9)	35 (26.7)	6 (4.6)	1 (0.8)
South Western	16 (25.0)	25 (39.1)	22 (34.4)	1 (1.6)	0
Southern	57 (43.5)	39 (29.8)	23 (17.6)	8 (6.1)	4 (3.1)
West Midland	13 (26.0)	14 (28.0)	21 (42.0)	2 (4.0)	0
Western	33 (35.1)	37 (39.4)	23 (24.5)	1 (1.1)	0
Yorkshire	25 (25.8)	25 (25.8)	22 (22.7)	13 (13.4)	12 (12.4)
Wales	107 (34.3)	116 (37.2)	73 (23.4)	13 (4.2)	3 (1.0)

Percentages are in parentheses.

Table 7.6. Convenience of layout, by Advisory Centre Provinces in England and Wales: responses to question B2 on Primary Return ('Is farm conveniently laid out?').

Province	Response			
	Yes	Moderately	No	No response
East Midland	76 (66.1)	31 (27.0)	8 (7.0)	0
Eastern	88 (52.7)	55 (32.9)	23 (13.8)	1 (0.6)
North Western	17 (54.8)	14 (45.2)	0	0
Northern	31 (38.3)	37 (45.7)	13 (16.0)	0
South Eastern	76 (66.1)	27 (23.5)	12 (10.4)	0
South Western	32 (56.1)	18 (31.6)	7 (12.3)	0
Southern	72 (61.5)	35 (29.9)	8 (6.8)	2 (1.7)
West Midland	59 (89.4)	4 (6.1)	3 (4.5)	0
Western	35 (40.7)	32 (37.2)	17 (19.8)	2 (2.3)
Yorkshire	64 (67.4)	18 (18.9)	13 (13.7)	0
Wales	118 (38.7)	110 (36.1)	75 (24.6)	2 (0.7)
England and Wales	668 (54.0)	381 (30.9)	179 (14.5)	7 (0.6)

Percentages are in parentheses.

the 1946 Summary as 'Good' (54%), 'Fair' (33%) and 'Bad' (13%) (MAF, 1946: 35).

Farm fragmentation and layout was also the main theme in at least one of the early post-war publications making use of the survey material. James Wyllie, Provincial Agricultural Economist at Wye College during the war and through to 1951, investigated the theme for the South Eastern Province (Wyllie, 1946).[33] Farm layouts, multiple holdings and systems of tenure were the foci of his 1946 publication. The second of these broadly corresponded with fragmented holdings as herein studied, with the definition being adopted of the same farmer running more than one holding as a farm business. On this basis Wyllie found 5.7% of the holdings in Wye Province to be multiple holdings, although they accounted for as much as 32.6% of the crops and grass. The NFS Summary, using data from 24 counties, suggested that about 70,000 holdings – nearly one in four – was subject to some degree of severence (Whitby, 1946: 337; Wyllie, 1946: 20).

Farm fragmentation in the sense of physical separation could arise from a well-developed characteristic of livestock farming systems, namely the right to participate in common grazings, which might entail the movement of stock over considerable distances. Details of such grazing rights were returned in the Primary Return as a reply to the questions generally relating to tenure: 'Has farmer grazing rights over land not occupied by him?' and, 'If so, nature of such rights'. The extent to which such rights existed is, of course, very much a function of the agrarian histories of the differing localities chosen for the present sample, and it was felt that they could in no way be generalized to give an indication of the spatial variation by Province of such rights. The 1946 Summary referred to 23,500 occupiers (8%

(A)

Ashburnham hereditament numbers 1910

The smaller hereditaments with their numbers
have been omitted from this sequence.

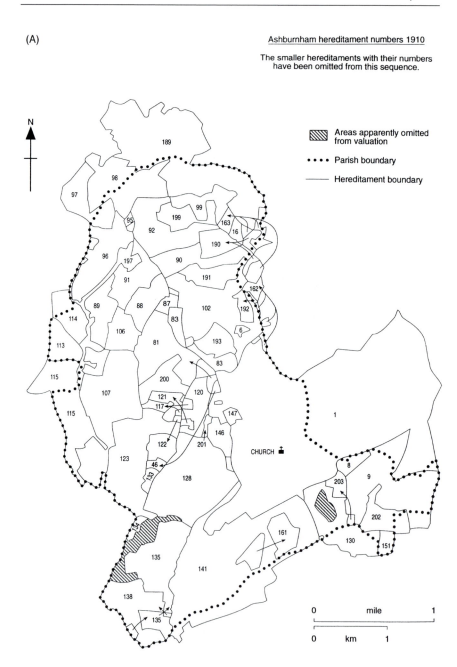

Fig. 7.1. Ashburnham parish, East Sussex: a comparison of farm boundaries *c.* 1910 (A) and 1941–1943 (B) (*opposite page*) (source: PRO IR 58/29198-29200; MAF 73/41/42-3, 56-7).

(B)

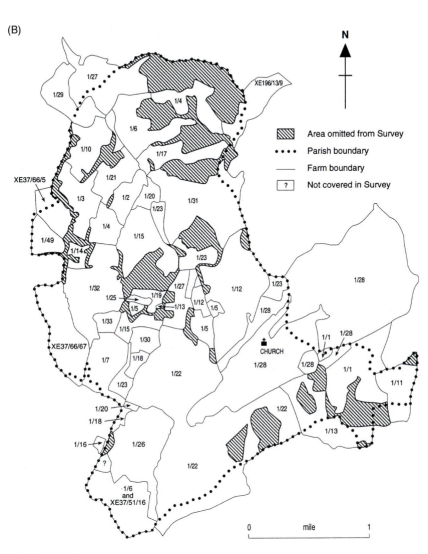

Fig. 7.1. (cont.)

of the total number) with grazings rights, which took one of three forms: grazing rights over common land, including hill grazing; land carrying stock agisted by the occupier; and land seasonally occupied such as that used for summer grazings (MAF, 1946: 15). The present analysis found 19% of the holdings in the sample to have grazing rights, but the inclusion in the national sample of parishes such as Denny Lodge in Hampshire with access to the New Forest, Shaugh Prior on the edge of Dartmoor, or New Romney in Kent, with grazings over an adjoining cricket field and golf course, would have necessarily entailed greater possibilities of grazing rights. In many cases commoners handed over their land to the Ministry to be used as required, but the answers given as to the nature of such rights do give an interesting insight into local farming systems (Easterbrook, 1943: 36). There are references to several acres of 'grasskeeping' let by individual owners in the New Forest; to '100 sheep gates on Nether Moor' in the East Midlands; to the use of '10 acres of Bank. Great Ouse Catchment Board, Cambridge'; to 'unlimited rights Townbank Kinniside' in Cumberland; or 'Rights on Clydach Mountain. 48 ewes, 7 milking cows, 4 heifers' from South Wales.

With the available information, it is clearly possible to reconstruct many facets of farming during the Second World War. However, it is also possible to compare the situation as revealed in the National Farm Survey 1941–1943 with an earlier, or indeed later, period. To illustrate this, the parish of Ashburnham, East Sussex, was taken as a case study to examine changes in farm structure between 1910 and 1941–1943. The 1910 picture is drawn from the records of the Finance (1909–10) Act 1910 (Short, 1997: 200–201).[34] A direct comparison can then be effected using the relevant map sheets for the National Farm Survey to assess the extent to which the First World War and the post-war depression affected farm boundaries in this parish of tenanted farms, nearly all belonging to the Ashburnham estate, centred on Ashburnham Place, which is located next to the church (Fig. 7.1). Remembering that the 1941–1943 survey excluded woodland, which was included in 1910, there are otherwise many similarities in the two maps. Most of the farm boundaries in the west and north of the parish remain the same, but there are changes to the area around the house and church, which involves the partitioning of the parkland by 1941–1943, and the expansion of the large southern farm referred to as 1/22 in 1941–1943 – in fact Kitchenham Farm, which was the home farm for the Ashburnham family. The amount of fragmentation (indicated in Fig. 7.1 by linking arrows) has decreased in the period. Both maps also demonstrate the non-coincidence of the farm boundaries with the parish boundaries, an issue explored by Coppock for the Chilterns in the 1950s, and whose work involved the obtaining of access to the NFS maps, but who did not use the other documents (Coppock, 1955, 1960a,b,c; Godfrey, 1999).[35]

Farm Mechanization

The war invoked a massive upsurge in the use of machinery on British farms. The substitution of the tractor for the horse and the use of combine harvesters rather than binders and stationary threshing machines were (eventually) accepted as vital to the war effort. On the heavy Essex clays a tractor could plough 3 or 4 acres per day, compared with the 1 acre by a horse team, and timing and weather considerations were also rendered less important. Crucially also, the dependence on imported oil for the tractors was thought to be a price well paid when a further 2–3 million acres of land previously used for horse feed could now be ploughed (Martin, 1992: 124–139). On the day that war broke out there were over 50,000 tractors at work; this number had doubled by March 1942, and reached 203,000 by the beginning of 1946. There were similar increases in the uptake of tractor ploughs, disc harrows, drills, potato diggers, milking machines, etc. About 2 million hp had been added to the land, making British agriculture the most highly mechanized in Europe (Easterbrook, 1943: 34).

As an essential component of modernity in the countryside, the diffusion of mechanization and the wider context of the relationship of farming to the scientific and engineering interests is clearly touched on by the NFS. The detailed examination of individual farms in respect of their machinery and electrification certainly demonstrates the interest of those who constructed the survey, and the broad convergence of farming and science is an acknowledged outcome of the wartime years, building on a more gradually mounting acceptance by farmers prior to the war (Whetham, 1978: 128). The experience gained in 1939–1945, the knowledge acquired from the machinery pools of the County War Agricultural Executive Committees (CWAECs) (Fig. 7.2) and from those attached to the committees (such as the

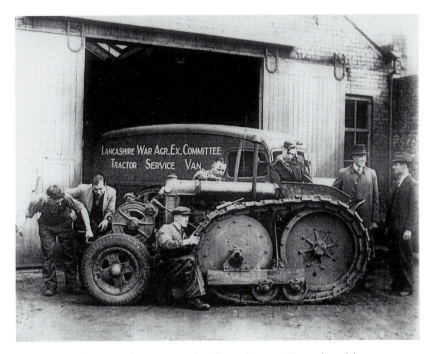

Fig. 7.2. Lancashire WAEC tractor service depot (source: Reproduced from S. Foreman, *Loaves and Fishes* (HMSO, 1989) by permission of the Ministry of Agriculture, Fisheries and Food).

staffs of the agricultural colleges that had been closed down in 1940), and the use of the CWAEC machinery in the post-war years all provided the essential context of the great post-war scientific revolution in British farming. A triumphalist wartime demonstration of mechanized farming using diesel-powered bulldozers is described by J. Wentworth Day (1943).

Some of the most powerful images of wartime agriculture are those of large-scale mechanization implying invincible mechanical progress (Fig. 7.3). Rapid reclamation of waste lands required large numbers of new machines. One of the best known and most publicized schemes was the Dolfor Scheme in Montgomeryshire (Ministry of Information, 1945b). Here

Fig. 7.3. Women's Land Army spearheading mechanized farming on the South Downs. Harvesting at Toy Farm, Beddingham, Sussex, 1941 (source: East Sussex Record Office AMS 5666/1/12).

Montgomeryshire WAEC ran a large-scale land reclamation scheme using the first batch of heavy caterpillar track-laying vehicles delivered through the lend–lease scheme. They were HD7s, the biggest used in Europe at that time, as well as smaller Fordsons and Fergusons. About 800 acres of upland bracken was ploughed up, with men working 80 man hours a day for a couple of months. They turned this area into the 'biggest potato patch in British history', ploughing up to 335 m above sea level – well above the normal cultivation level. The Montgomeryshire WAEC had been running a small experimental scheme with a few small tractors, then MAF took it over and introduced the American crawler tractors. They were the first people to plough 1 acre per hour per man in the country. The scheme lasted about 3 months.[36]

The striking images of mechanical land reclamation perhaps obscure a quiet revolution taking place on farms across the country. Although land reclamation took place in most counties, this revolution was the replacement of the horse by the tractor for the day-to-day farming activities of ploughing and harvesting. Data from the NFS should allow us to answer some basic questions about this change. When did the tractor replace the horse? On what sort of farms did the tractor first appear? Were there regional variations in the uptake of mechanization? The annual Census Returns available to MAF gave no information about mechanization and so the advent of the NFS provided a welcome opportunity for MAF to measure the availability and use of agricultural machinery.

The importance given to farm mechanization by MAF can be judged by the effort that went into collecting special information on the Supplementary Form. Yet, despite the fact that this form was designed, sent, sorted, copied and stored specifically as part of the NFS, no analysis of the mechanization material is included in the 1946 Summary Report. The reasons for this have not become apparent from our investigation of the minutes and committee papers associated with the writing of the Summary Report. However, the results of our analysis of data from our three samples suggest that the principal reason may have been the relatively poor quality of the data. This poor quality may be due to the novelty of the form (farmers had not been questioned on mechanization in the Census before), by the fact that the form is poorly designed with ambiguous questions and by the lack of room for farmers to provide adequate responses.

The Supplementary Form does not fully represent the motive power available within a parish, district or county, as the CWAECs held many tractors and other machinery themselves which were leased out or otherwise contractually engaged to plough up land within their jurisdiction. These pools of motive power were very significant in the day-to-day implementation of the Plough-up Campaign and the works overseen by the CWAECs across the country. They were also important for high-profile schemes such as the Dolfor Scheme in Montgomeryshire. The Supplementary Form does, however, provide a wealth of information about the use of machines on farms at a time when mechanization was going through a period of accelerated change.

In this section we evaluate the quality of the mechanization data. The two main categories in the section on motive power on the Supplementary Form were 'Fixed or Portable Engines (excluding Motor Tractors)' and 'Tractors'. Farmers were asked to provide information on the number and horse power of the following engines: water wheels or turbines in present use; water wheels not in use, but easily repairable; steam engines; gas engines; oil or petrol engines; electric motors; and other engines. For tractors, farmers had to specify the number, horse power and make or model of tractors on the holding. They were asked to distinguish between wheel tractors for field work, wheel tractors for stationary work only, and track-laying tractors. The data in the sections below ('Fixed and portable engines' and 'Tractors') give some idea of that degree of consistency, as well as the range of responses that were given. Where a sufficient number of holdings returned information on engines of some kind, they have been given as a percentage of the total number of holdings for that sample. Where figures were given for horse power, these were also manipulated to produce average figures.

Fixed and portable engines

Most responses in this section relate to 'Oil or Petrol Engines' and to 'Electric Motors'. Answers are given very infrequently in the other categories, although examples of all were found in our samples. Horse power (hp) details are very often not given, or the hp of one engine only is given when there are two or more on the holding. Sometimes details of the make and use of the engines and motors are given. Engines for pumping and working elevator gear are relatively common, as are electric motors for milking machines. Lawn mowers are frequently entered in this section as well.

Very few farms returned information on water wheels. In the *National Sample* only ten holdings had working water wheels. They were all in Wales, apart from one in Devon, and the maximum hp was 12. Only two holdings had water wheels that were currently out of use but repairable – one in the North Riding of Yorkshire, the other in Cardigan. In the *Sussex Sample* nine holdings had water wheels, with a mean hp of 7.75. There were no water wheels not in use but repairable. There was only one water wheel in the *Midlands Sample* and this not in use but repairable.

Steam engines were extremely rare. In the *National Sample* five holdings returned steam engines – one in Monmouthshire, one in Cornwall, one in Lincolnshire (Lindsey) and two in West Sussex. Their mean hp was 6.8. In the *Sussex Sample* four holdings had steam engines, one of these having three different engines. There were no steam engines returned within the *Midlands Sample.*

Gas engines were also uncommon. Only four such engines were returned in the *National Sample*: one in Herefordshire, two in Cornwall and one in West Sussex. In the *Sussex Sample* ten holdings had gas engines and there was only one in the *Midlands Sample.*

The most common return was for oil and petrol engines. In the *National Sample*, 364 holdings (27.4%) returned one or more oil or petrol engines. Most had a single engine but 95 had two or more oil and petrol engines. The average hp was 4. The holdings with more engines did tend to be larger, although some smaller, more intensive horticultural holdings had more than one engine. In the *Sussex Sample*, 415 holdings (34.9%) returned one or more oil or petrol engines, while the equivalent figure for the *Midlands Sample* was 103 (21.6%).

Electric motors were also reasonably common. In the *National Sample*, 39 holdings returned electric motors. Their mean hp was 4, although some were up to 30 hp. The holdings with more engines did tend to be larger, but one holding with seven electric engines was only 5 acres, so there were anomalies. In the *Sussex Sample*, 99 holdings (8.4%) had electric motors, while in the *Midlands Sample* the equivalent figure was 31 (6.5%).

In the *National Sample*, 31 holdings returned some kind of entry under 'Other' kinds of engines; in the *Sussex Sample* the equivalent figure is 39 and in the *Midlands Sample* it is nine. A very wide range of vehicles and engines, often adapted for agricultural use, is specified. Examples include 'Austin Car Auto Culto' and 'Sunbeam converted to lorry' (both in Hampshire); 'Blackstone crude oil barn engine' (Dorset); 'wood sawing plant' (Cumberland); 'Concrete mixed [*sic*] Two fruit sprayers' (Surrey); 'Allen motor Scythe' (Wiltshire); 'Lister Crude Oil' (Herefordshire); 'Morris car (used for hay sweep)' (Sussex); and 'Converted Humber car' (Warwickshire).

The total number of engines of all kinds, and horse power, was calculated for all holdings in the three samples. In the *National Sample* the maximum number of engines held by any one holding was 17. The mean hp present on holdings that had some form of engine was 6.7; for all holdings it was 2.1. The mean hp per acre of holdings with some form of engine was 0.12; for all holdings it was 0.04. The holdings with the greater number of engines and greater hp per acre did tend to be larger than average, but this was not always the case. In the *Sussex Sample* the maximum number of engines held on any one holding was 14. This particular holding also returned the largest area of land producing horticultural crops. However, some large horticultural holdings did not return any engines. In the *Midlands Sample* the maximum number of engines held by any one holding was 13. The mean hp present on holdings that had some form of engine was 8.0; for all holdings it was 2.2. The mean hp per acre of holdings with some form of engine was 0.10; for all holdings it was 0.03.

Overall the NFS shows that fixed and portable engines were important on many types of holding. Clearly, by the early 1940s, steam engines had virtually disappeared. As the mid-century is often thought of as the Golden Age of the petrol and oil stationary engine (illustrious names such as Villiers and Lister spring to mind), it is perhaps surprising that only just over a quarter of the holdings in our *National Sample* had such an engine. Clearly some types of holding, especially those with fruit or horticultural enterprises, were more likely to make use of portable engines than others. Electric motors had yet to make a big impression on farm motive power.

Tractors

The most common category of tractor returned was 'Wheel Tractors for fieldwork'. Several manufacturers are named: the most popular make is Fordson, but International, John Deere and Allis Chalmers also occur relatively frequently. There are many cases of cars and lorries that have been converted for tractor work – Austins, Morrises and even a Studebaker. The Fordsons vary between 14 hp and 35 hp. There is often uncertainty about the horse power of the tractor. There are several instances of the question 'Number in Figures' being answered as the engine number of the tractor, or, in one case, the licence number. Cletrac and Caterpillar are the most common makes of track-laying tractors.

Just over one-fifth (278; 21%) of the holdings in the *National Sample* returned one or more field tractors. Of these, 236 had one, 32 had two, seven had three, and four had four or more (one in West Sussex had 11). The average hp was 21, with a range of one to 40 and a modal value of 20, but hp information was missing from many (66; 24%) forms. Seven holdings did not give any make, but of the 344 tractors returned, the majority were made up of variants of the following names: Fordson (173), International (35), Case (13), Ferguson (12), Allis Chalmers (nine), Massey (seven), John Deere (seven), David Brown (six) and Auto Culto (five). There was also a number of miscellaneous converted cars and lorries.

The *Sussex Sample* had rather a higher proportion (433; 36.5%) of holdings with one or more wheeled field tractors. Of these, 312 had one, 86 had two, 18 had three, and 13 had four or more tractors. Over a quarter (116; 27%) of these returns provided no hp. The average hp was 20, with a range of 1.5 to 40 and a modal value of 20. The most common make by far was Fordson (325) followed by International (62), Allis Chalmers (24), Simar Rototiller (eight), Austin (five), Case (four), Ferguson (three), Rushton (three) and David Brown (one). There were a number of miscellaneous converted cars and lorries and a 'Trusty' specified as a 'milking tractor'.

Just less than a fifth of the *Midlands Sample* holdings (94; 19.7%) returned one or more wheeled field tractors. Most (75) had one tractor, 15 had two, three had three and one had four or more. The average hp was 15.5, with a range of 5.5–30 and a modal value of 20. Only a single holding did not give any make; the majority were Fordson, with 'Land', 'Green Spot', 'Agricultural' and 'Utility' given as model names (71), Massey [Harris] (seven), Case (seven), International (six), Ferguson (two), Allis Chalmers (one) and David Brown (one). Again there was a 5.5 hp tractor called a 'Trusty' and several converted cars.

Very few holdings used 'Wheeled tractors for stationary work only'. In the *National Sample* five holdings returned a single stationary wheel tractor. Two of the holdings were in Norfolk. The makes were given as International Industrial, Overtime, Titan, Austin (converted Austin car) and International. In the *Sussex Sample*, 11 holdings had at least a single stationary wheel tractor. Three were Fordsons, three were International, and there was one each of International Junior, Titan 10/20, Rushton, Overtime and Cletrac. The two holdings in the *Midlands Sample* with stationary tractors gave the makes as 'Converted Car' and 'Trusty with Austin 7 engine'.

Track-laying tractors were much less common than wheeled tractors. In the *National Sample* only six holdings (0.48%) had a track-laying tractor. The mean hp was 17.6. The makes were Caterpillar, Bristol, Ransoms, Allis Chalmers and MW Chalmers D4. The *Sussex Sample* produced rather more holdings (23; 2%) with one or more track-laying tractors. Most of the parishes where track-laying tractors were found were on the Downs.[37] The makes were five Caterpillar D2, one Caterpillar R2, two Caterpillars D2 Diesels (all the Caterpillar tractors were either 22, 25 or 30 hp), two Bristol, one Bristol Farm Tractor, one Allis Chalmers, three Cletrac, one Fowler, one International (Diesel), one International, two MW Chalmers D4, one Ransomes and one Ransomes Garden Cultivator (both the Ransomes were 6 hp). Seven of the holdings were under 50 acres or did not return an acreage on the Primary Return and they tended to return the Bristol and Ransomes tractors, whereas the others were between 174 and 2482 acres (with nine being over 600 acres). The bigger holdings tended to have the Caterpillar, Cletrac or International tractors. In the *Midlands Sample* four holdings (0.9%) returned one or more track-laying tractors. The makes were two Caterpillar, one Caterpillar 22, one Caterpillar R2, and one Ransome Garden Tractor. The holdings were between 320 and 542 acres in size.

Did the fifth of all holdings in the *National Sample* that had tractors have any special characteristics? A brief examination of the data shows that holdings with tractors were consistently larger than holdings generally, with a mean acreage of 181 acres as compared with 124 acres. In terms of rental value of holdings held by tenants, those with tractors had a mean of £192 as compared with mean for all holdings with the relevant data available of

Table 7.7. The percentage of holders with different gradings in the *National Sample* with tractors compared with the percentage of all holders.

Grading	Holders with tractors (%)	All holders (%)
A	64	54
B	30	38
C	6	7
No grading given	0	1

£108. In terms of rental value per acre, however, holdings with tractors had a mean value of £1.4 compared with a mean for all holdings with the relevant data available of £1.6.

Perhaps surprisingly, there was no relationship between the tenure of a holding and the ownership of a tractor. A quarter of all holders with tractors were owner-occupiers, 63% were tenants and 13% were both owner-occupiers and tenants. This compares with figures of 28%, 62% and 10% for all holders. Tractor owners were slightly more likely to be graded A than B. Ownership of a tractor did not, however, preclude being classed C (see Table 7.7).

Analysis of the labour figures shows that holdings with tractors had more labourers per acre (0.038) than holdings that did not (0.028). In terms of different types of labourers, holdings with tractors had a mean of 0.03 whole-time workers per acre, as compared with all holdings which had a mean of 0.19 whole-time workers per acre. Overall, therefore, farms that had tractors were likely to be larger in area, employ more labour and achieve a higher grade than farms that did not. There is also a faint indication of the eventual effect of tractors on the horse population: the average number of horses per acre for holdings with tractors was 0.032 compared with 0.041 for all holdings.

Conclusion

This examination of information concerning stationary engines and tractors confirms that the NFS records hold an enormous amount of information on farm machinery. Clearly the NFS will be the starting point for any study wishing to chart the rise of mechanized agriculture in England and Wales in the mid-20th century. The NFS confirms that only a fifth of all farms had tractors in the early years of the war, and that it was generally the larger farms that had tractors earliest. Fordson was by far the most popular manufacturer. The higher proportion of farms with tractors in Sussex confirms that there were strong regional differences in wartime mechanization, a subject that demands further research.

Fig. 7.4. Women join the farm workforce: original coloured artwork on cardboard, with on reverse, 'A Agricultura en Inglaterra – Sir John Russell' (source: PRO INF 3/1743, is Crown copyright and is reproduced with the permission of the Controller of Her Majesty's Stationery Office).

Women in the Records of the National Farm Survey

The dominant image of wartime women in agriculture is that of the Land Army girl (Figs 7.4 and 7.5). There is no doubt that the Women's Land Army (WLA) was of enormous importance for the war effort, and had widespread social effects. There is little information in the NFS records about the WLA, but what the records do provide is the opportunity to examine the role of women in agriculture on a day-to-day level. The role of women in farm families has received some attention in recent years (Gasson, 1980; Whatmore, 1991). What the NFS enables us to do, for the first time, is to assess the number of female farmers, and to examine the role of female labour in agriculture. The NFS records provide an invaluable threshold to explore the role of women in British agriculture in the 20th century. As a

Fig. 7.5. Women's Land Army members working in a potato field (source: PRO MAF 59/145, is Crown copyright and is reproduced with the permission of the Controller of Her Majesty's Stationery Office).

means of evaluating the data found in the NFS, our database was searched in various ways to see if women could be located as holders of farms, owners of land or labourers on the land.

Identification of women farmers

What information does the NFS provide about the number and proportion of the holders of farms who were women? When the data samples were collected it was noted that in the top section of the Primary Return, where the details of the name of the farm, name of farmer, address and so forth were collected, a gender distinction was available that did not have to be based on guesswork such as the interpretation of Christian names. This distinction was in the category 'Name Farmer'. If the holder was male, then merely the name would be given. Occasionally titles such as Mr, Capt., Lord or Earl might be present; however, it was a norm of the Survey that the default value for a holder's gender was assumed to be male. If the holder was female then a title such as 'Miss', 'Mrs' or 'the Misses' would be entered. In the few cases where there was no Primary Return the name of the holder could usually be gleaned from the addressograph. This generally had titles of both genders and so provides an alternative way of locating women holders. We stored this as additional information in our database.

Therefore, female holders in our database could be found by searching the relevant variables for the categories 'miss', 'mrs', 'lady', 'viscountess' and so forth. This did turn up some non-applicable records, as in cases where these titles referred to previous holders or where words such as 'missing' were present. Once these misleading records had been weeded out, holders who could be identified as female with a high degree of confidence were included in the database. The results, shown in Table 7.8, indicate that 80 (6%) of holders in the *National Sample* were women and that a similar proportion of holders were female in both the *Sussex Sample* (90; 7.6%) and *Midlands Sample* (33; 6.9%).

Female holders were substantially more likely to be owner-occupiers than all holders in all three samples. In the *National Sample*, for example, well over half (58%) of the female holders were owner-occupiers, compared with under a third (28%) of all holders. In the *Sussex Sample* the proportion of owner-occupiers rose to almost two-thirds (64%) of female holders (Table 7.8). Female holders were rather less likely than holders generally to be classed as full time, and more likely to be classed as part time or as a hobby farmer (Table 7.8). The temptation to infer that most female holders did not farm on their own account should be avoided, as over two-thirds of the women holders in the *National Sample* were classed as full time.

There was some regional variation in the proportion of female holders who also had other occupations, but these did not differ significantly from all holders in the samples (Table 7.8). In the *National Sample* a number of other occupations were specifically gendered: the Hon. Mrs B. Jones was given as 'Housewife'; Miss M.E. Sands as 'Housekeeper to mother' (she was also given as the holder of two other holdings); Mrs E. Wilkins as 'Gentlewoman of independent means'.[38] In the *Sussex Sample* a number of female holders were noted as racehorse stable proprietors, and Miss Hett was given as 'A lady of leisure'.[39]

Grading of women farmers

The grades given to farmers under section D of the Primary Return were analysed to see if women farmers were classed differently from farmers as

Table 7.8. Female landholders, tenure and type of occupation.

	National Sample	Sussex Sample	Midlands Sample
No. of holdings in sample	1330	1186	477
No. of holdings held by females	80 (6.0)[a]	90 (7.6)[a]	33 (6.9)[a]
No. of holdings with tenure details available	1226	1055	410
No. of holdings with tenure details available occupied by females	74	84	27
Female owner-occupiers	43 (58.1)[b] (27.7)[c]	54 (64.3)[b] (33.3)[c]	9 (33.3)[b] (22.7)[c]
Female tenants	27 (36.5)[b] (62.2)[c]	24 (28.6)[b] (58.4)[c]	14 (51.9)[b] (66.0)[c]
Female joint tenants/ owner-occupiers	4 (5.4)[b] (10.0)[c]	6 (7.1)[b] (8.3)[c]	4 (14.8)[b] (11.5)[c]
Female full-time holders	50 (67.6)[d] (75.7)[e]	46 (54.8)[d] (69.8)[e]	16 (59.3)[d] (66.8)[e]
Female part-time holders	6 (8.1)[d] (11.2)[e]	4 (4.8)[d] (12.5)[e]	5 (18.5)[d] (13.4)[e]
Female spare-time holders	5 (6.8)[d] (8.3)[e]	6 (7.1)[d] (5.1)[e]	4 (14.8)[d] (11.5)[e]
Female hobby farmers	7 (9.5)[d] (2.2)[e]	14 (16.7)[d] (7.3)[e]	2 (7.4)[d] (2.0)[e]
Other female holders	6 (8.1)[d] (2.6)[e]	14 (16.7)[d] (5.3)[e]	0 (0.0)[d] (6.3)[e]
All holders with other occupations	290 (23.7)[f]	233 (22.1)[f]	141 (34.4)[f]
Female holders with other occupations	18 (24.3)[g]	16 (19.0)[g]	10 (37.0)[g]

[a]   Percentage holdings with female landholders.
[b]   Percentage totals for female tenure.
[c]   Percentage totals for tenure of all landholders in the Samples.
[d]   Percentage totals for female holders.
[e]   Percentage totals for all holders in the Samples.
[f]   Percentages of total no. of holdings with tenure details available.
[g]   Percentages of total no. of holdings with tenure details available occupied by females.

a whole. The proportions of A, B and C farmers were calculated for the three samples as a whole and for the holders identified as being female within them (Table 7.9). Although our samples are small, it is interesting to note that female farmers are not graded A as often as farmers generally, and were more likely to be given the lower grades in all three regions. The comments given when personal failings were identified for female farmers were not noticeably different from the kind given for farmers as a whole, in any

Table 7.9. Grading of holders within the National Sample and the grading of holders identified as female for comparison.

Grades	Number of holders	Percentage of holders	Number of female holders	Percentage of total female holders
A	651	53.1	28	37.8
A−	5	0.4	0	0.0
B+	42	3.4	3	4.1
B	412	33.6	32	43.2
B−	16	1.3	2	2.7
C+	0	0.0	0	0.0
C	89	7.3	7	9.5
C−	1	0.1	0	0.0
No grading given	10	0.8	2	2.7
Total	1226		74	

of the three samples, and only a few could be said to be gender related, such as for Mrs Elizabeth Parry who was graded B− and where the comment 'Widow' is given as a 'Personal Failing', and for Mrs Harriet V. Eberhardt, graded C, where the comment 'Lack of interest and strength' is recorded.[40]

Under 'General Comments' (Section E of the Primary Return) the comments for women are not generally different from those for farmers as a whole, and in fact on several occasions female farmers are congratulated on their management of holdings. One peculiar comment in Nottinghamshire is as follows: 'Mr Maltby is the farmer. Mrs Maltby stays indoors but decides farming matters. Lack of education and executive ability'.[41] Seven of the holdings held by women in the *National Sample*, and one in the *Midlands Sample*, seemed to be hobby farms or parkland which was not being developed for agricultural purposes, and was for the grazing of hunters or deer. A further two in Sussex were described as 'a small pleasure farm' or 'a comfortable country house'. There is some indication of subtle distinctions in social status among female farmers, which could be pursued further in examining the ownership of land, the tenure of female holders and the size of farms held by female holders.

Size of farm

Various measurements of average figures for the size of farm, rental values and rent per acre for holdings occupied by female holders as compared with all holders were obtained for the three samples (Table 7.10a, b, c). The figures show that holdings occupied by women were smaller than those of all holders in each sample. This difference was particularly noticeable in the *Sussex Sample* where women's holdings were on average about half the size of those for all holders. Women also occupied holdings with a lower than average gross rental value and a lower rental value per acre than all holders.

Women landowners

Women are also present in the NFS records as landowners; in fact, they occur in this role more frequently than as holders. In the *National Sample*, 16% of holdings were owned entirely or in part by female landowners. In the *Sussex Sample* the figure was also 16%, and in the *Midlands Sample* it was 12%. The female landowners owning the largest acreages would usually also be titled in some way, such as Lady Curre in Itton, Monmouth; Lady Gerald Wellesley in Withyham, East Sussex; Lady Catherine Ashburnham in Ashburnham, East Sussex; or the Hon. Mrs Methuen-Campbell in Rhossili, Glamorgan. The majority of female landowners, however, were simply Mrs or Miss and would own part of or one or two holdings in a parish. Although acreage figures or OS parcel numbers were sometimes given for the owners returned under Section A2 of the Primary Return, this was not done in any consistent way and so it is difficult to discover with much certainty the spatial extent or acreage of ownership for any individual – male or female – without making use of the NFS maps.

Women as labourers in the National Farm Survey

The proportion of female labourers was calculated for holdings in all three samples. In the *National Sample* just over half (702; 55%) of the 1275 holdings returned some labour under questions 35–42 of the Census Return. Of the holdings with labour, as many as a quarter had some female labour and three-quarters did not. For all 1275 holdings, 175 (14%) returned some female labour. Where there was some female labour on a holding, on average it comprised 10% of the total labour on that holding. Out of a total of 2339 labourers returned, 321 (14%) were female. Of the 321 female labourers, 205 (64%) were whole-time workers and 116 (36%) were casual workers. These figures can be compared with those for all the 2339 workers returned, of which 1915 (82%) were whole-time and 424 (18%) were casual. Female labourers were therefore 11% of all whole-time workers and 27% of all casual workers.

In the *Sussex Sample* the answers to the labour questions from the Census Return show that over two-thirds (768; 69.5%) of the 1105 holdings returned some labour. Of the holdings with labour, 220 (29%) had some

Table 7.10. Comparison of mean rental values, acreages and rent per acre for all holders and female holders.

(a) National Sample.

	Rent (£)		Acreage		Rent per acre	
	All holders	Female holders	All holders	Female holders	All holders	Female holders
Mean	108.0	89.6	124.0	100.7	1.6	1.5
Median	75.0	42.0	75.0	35.0	1.0	1.2
Mode	60	6	18	8	2	2

(b) Sussex Sample.

	Rent (£)		Acreage		Rent per acre	
	All holders	Female holders	All holders	Female holders	All holders	Female holders
Mean	143.2	100.9	166.6	89.3	1.8	1.6
Median	91.3	90.0	74.0	82.0	1.1	1.0
Mode	40	N/A	95	N/A	1	N/A

(c) Midlands Sample.

	Rent (£)		Acreage		Rent per acre	
	All holders	Female holders	All holders	Female holders	All holders	Female holders
Mean	116.0	107.2	89.8	85.6	2.2	1.5
Median	84.8	80.0	57.8	57.5	1.5	1.5
Mode	55	80	18	57.5	1.25	N/A

female labour and a fifth of all holdings (220; 20%) returned some female labour. Where there was some female labour on a holding, on average it comprised 37% of the total labour on that holding. Out of a total of 3845 labourers returned, 554 (14%) were female. Of the 554 female labourers, 340 (61%) were whole-time workers and 214 (39%) were casual workers. These figures can be compared with those for all the 3845 workers returned, of which 3320 (86%) were whole-time and 525 (14%) were casual. Female labourers were therefore 10% of all whole-time workers and 41% of all casual workers.

In the *Midlands Sample* just over half (241; 53%) of the 455 holdings had some labour. Of the holdings with labour, 53 (22%) returned some female labour. Where there was some female labour on a holding, on average it comprised 47% of the total labour on that holding. Out of a total of 871 labourers returned, 143 (16%) were female. Of the 143 female labourers, 61 (43%) were whole-time workers and 82 (57%) were casual workers. These figures can be compared with those for all 871 workers returned in the sample, of which 697 (80%) were whole-time and 174 (20%) were casual. Female labourers therefore comprised 9% of all whole-time workers and 47% of all casual workers.

These results show that, overall, between a quarter and a fifth of all holdings that employed labour employed some female labour. They also show how the NFS records can be used to show regional variations in the use of female labour on farms. For example, in the *Midlands Sample* and *Sussex Sample*, over 40% of casual workers were female, whereas the figure for the *National Sample* was 27%. Nationally, when females were employed on a farm they on average formed 10% of the labour force, yet for the *Midlands Sample* the equivalent figure was 47%. Clearly a whole range of factors, including size of farm and number of employees, will affect the interpretation of these figures. Nevertheless, these examples show the potential value of the NFS in looking at trends in female labour. It would be interesting to explore the extent to which Land Army girls were returned with other female labour on the Census Form.

Women in the administration of the National Farm Survey

Almost all the people named in the administrative documents of the NFS are men. There are, nevertheless, a few interesting cases where women are mentioned, and some specific gender distinctions are made. In a memorandum concerning requirements for office staff to work on the NFS, for example, it was noted: 'For internal staff, he would require an experienced indexer in charge of the records – preferably a woman trained in library work and filing.'[42] A note from Kendall to Black (30 May 1941) provides the names of men who might be suitable for the NFS, who were considered the 'right type to work among farmers'. This list included Dr Ethel Mellor, DSc, a lecturer in education from Paris who 'might be useful in an office post'.[43]

Elsewhere in the files and the actual forms of the NFS there is a consistent gender bias, where the default values are always male and gender-specific pronouns used even when women are mentioned in the same sentence or paragraph. This reflects, of course, the accepted practice of the time. Despite not being considered suitable for work as Surveyors or as executive staff (there were one or two women Surveyors, but they were only a tiny minority) many women were involved in the NFS. Much of the copying of the forms was done by women, such as Miss Ann Nowill from Sutton Bonington, hired temporarily to work on the Survey, and who retired from Sutton Bonington after working for many years on the Farm Management Survey. She was almost completely anonymous with the NFS, as she did not sign or initial any of the forms she copied. There was a deliberate policy to employ women for jobs such as filing in order to keep down costs. Women from the Poultry Section were brought in to undertake the work on pages 2 and 3 of the 398/SS form at MAF's Statistical Branch at St Anne's. Women were hired almost exclusively to punch the cards for the Hollerith computers at Rothamsted used to produce the summary report.[44]

Conclusion

The NFS contains a great deal of information about the roles and position of women in farming, land ownership and agricultural labouring during the Second World War. This can be used to complement existing and emerging knowledge about the Women's Land Army, which is only infrequently present in the records of the NFS, despite its importance in addressing the labour shortage during the war. The data range from anecdotal observation under 'General Comments' to the statistics concerning female labour. The holders who were identified as female can be cross-tabulated with any of the other variables to discover if there was a general or regional profile of the type of holding occupied by women, as well as simply to describe the holdings they occupied.

However, the data must be treated with caution, as it is not even certain that all the female holders in the data set were identified. Two additional women in the *Sussex Sample* were found by chance when researching a completely different topic. Some of the holders identified as female were also somewhat ambiguous, as elsewhere in the records different names or titles were returned. Nevertheless, within the reasonably large sets of data that were collected, the presence of women in agriculture was found to exhibit certain consistent patterns both nationally and regionally which could not be discerned from any other source. For instance, the data indicates that approximately 7% of all holders were female, that they generally occupied smaller holdings of a lower rental value per acre than average, and that approximately 13% of all landowners identified in the NFS records were female. These findings are certainly new and indicate both general and specific uses of the NFS data.

Military Activities as Reflected in the Records of the National Farm Survey 1941–1943

As a theme running through the whole network of Ministry activities at this time, any reflection of militarization within the Survey merits careful attention. Stemming from the passing of the Emergency Powers (Defence) Act in August 1939, a series of defence regulations came into force, many of which affected the agricultural landscape through the intervention of the Air Ministry, the War Office and other departments. A great variety of records is therefore now available for more specialist researchers, and this section will limit itself purely to information on the activities of British and Allied forces, and their personnel, as well as on enemy action, which can be found in the NFS records – most commonly in the sections of the records set out below (Foot, 1999).[45]

The farm records with the greatest such information are those of the 'front-line' counties – Kent, Sussex, Surrey, Hampshire and East Anglia – although the records of inland counties generally also reflect the military need for training grounds, for airfields and for prisoner-of-war camps. This section is, however, based largely on an examination of the farm records of Wealden and Downland parishes of East and West Sussex, with further information being drawn from the *National Sample*. The theme was an emotive one, with the military taking over England's rolling (and productive) acres, and an intensely visual one (Figs 7.6 and 7.7).

The sources of information

The Primary Return

By far the most information on the effects of the military on the farms is given in the 'General Comments' section of the Primary Return. East Sussex records are particularly good here as the 'General Comments' are usually entered very fully. Some information on whether the farm owner or tenant is in the armed forces may be included in the 'Tenure' section of the form. Sometimes such information is also included in the 'Personal Failings' box of the 'Management' section. This latter section is occasionally used to state that the farm has been requisitioned by the military, and this information may be entered as well in the 'Conditions' section.

Fig. 7.6. Tanks in a Yorkshire harvest field (source: PRO INF 2/42 (TR156), is Crown copyright and is reproduced with the permission of the Controller of Her Majesty's Stationery Office).

Fig. 7.7. Barbed wire symbolism: the landscape of defence and farming (source: Rural History Centre, University of Reading, reproduced in S. Ward (1988) *War in the Countryside 1939–45*, p. 123).

The Census Return

Either page 1 or (more commonly) page 3 of the Census Return forms is often annotated 'Taken by the Military', or similar wording. Usually in these cases there is no Primary Return, and the Census Return (where it has been completed just ahead of the military requisitioning) serves as a record of the value of the agricultural land and livestock that were sacrificed for military needs. Very occasionally, the 'Labour' section of the page 3 form will include a reference to soldiers helping with the harvest.

The maps

The farm boundary maps often show military land use, and normally confirm the information given in the Primary or Census Return, although they usually supplement that information. The type of military use is generally written on the map face within the areas of the holdings affected, and the boundaries of the military requisitioned land are normally clearly indicated. In August 1941, for security reasons, the Ministry of Agriculture instructed the CWAECs not to show the military use of land on the maps. It asked the CWAEC cartographers to avoid the blank areas that might thereby appear on the maps, which themselves might indicate military areas, by using phrases on the map face such as 'Not used for agriculture'. It was even suggested that dummy farm codes be placed on the map to hide military use (see Chapter 3). A study of the East and West Sussex maps, however, shows that there is no obvious concealment of the military information, and they are marked very clearly with such labels as 'Military Area', 'Military Camp' or 'Aerodrome'.

The types of military information given

The date range of the military information from these sources could be almost anything within the years from 1940 to 1945, because the NFS was so long in completion, although it is most likely to relate to 1941–1943. Information annotated on the Census Return should generally show the situation as at 4 June 1941, but some of these annotations may have been made by Advisory Economists' staff when they were attempting to match the records at some time after June 1941. The dates of the information on the maps are even more uncertain. The maps were the last component of the total farm record to be completed, and this was sometimes not done until 1945 (or even later) and so the information may relate to any (or all) periods between 1940 and 1945. However, it seems likely that, as with the

Primary Return, the main period represented on the maps is 1941–1943, and that the military use of land shown is the situation at a date within that time span. This means that the great expansion in the military requisitioning of land, consequent upon the training, build-up and defences needed for the D-Day operations, will not be shown on the NFS maps; and neither will the defensive sites for the later operations against the V1s and V2s.

Despite these uncertainties, there are several themes that can be pursued. A selection is given here but there will be many more themes, especially when the NFS material is used in conjunction with other public records now becoming available (Cantwell, 1993).

Damage by Home Forces to farms and to agriculture

The principal information in the Primary Returns concerns damage done by the military, usually in connection with training needs, but sometimes as a result of their requisitioning of land for camps and headquarters and the erection of defence works. Thus at Patching, West Sussex, bushes were cut down by the military and 'thrown all about', and fences were destroyed, allowing rabbits to get into fields of wheat, damaging 9 acres.[46] Fences were also damaged in Alfriston, preventing sheep from being grazed, and defensive works in this area seriously interfered with the work. At Upper Beeding, 'numerous trenches' from the First World War were still inconveniencing a farmer, while at Buxted much of a farm's low-lying brookland had been spoilt by the construction of an anti-tank trench.[47]

Buxted suffered other deprivations: one farm was badly damaged by 1940 defence works, while another farmer suffered many losses in the summer of that year, including three cows that aborted. The compensation received is described as 'totally inadequate'. Defence works caused 'the maximum of inconvenience' to a farm in East Grinstead, and at Findon the military cut up the pastures. At West Firle, 'serious interference from the military' affected a farm's milking herd, and at nearby Berwick it was necessary to dispose of a flock of 180 ewes owing to 'the hill ground being made impossible by the military'. Similarly at Pyecombe, a Southdown flock had to be reduced owing to the barbed-wire entanglements erected by the Army on the rough grazing. The requisitioning of over half a farm at Falmer caused great difficulties to a farmer by the loss of farm buildings and much of the grazing land, and a farm at South Heighton was also 'much diminished by military occupation and disturbance' that had been going on since May 1940. Elsewhere, at St Just in Roseland (in Cornwall) a military camp occupied two fields on one farm, causing very heavy traffic on the farm roads.[48]

The requisitioning of farms and country houses

The requisitioning of farmland was undertaken in liaison with the relevant CWAEC and Land Commissioner. The requisitioning of large country houses, so famously described in *Brideshead Revisited* (Waugh, 1945), is not further dealt with here, although it should be noted that the Register of Accommodation maintained by the Ministry of Works (which had been started in 1938) contained entries for some 300,000 separate premises (Robinson, 1989: 5–13; Foot, 1999). There was frequently direct damage to the land as a result of military activity, and although there was provision for compensation, the very act of requisitioning rural land often meant a considerable loss to agriculture. Thus at Stoughton, 5 acres of permanent pasture were occupied by the military, while at Friston a farmer was left 350 acres out of 1000 (including the rough grazing). He stated, 'The Air Ministry and the Army have taken over 652 acres of my farm.' A Maresfield holding lost 49 acres to the military, and in Rodmell, 286 acres of one holding (including 92 acres of arable) were requisitioned. Kingston (near Lewes) also suffered from the requisitioning of agricultural land: 'Most of the rough grazing and some arable land on the hill have been taken by the military' and '250 acres have been taken recently by the military.' At Beddingham, land was taken for army training which had earlier been farmed by the East Sussex WAEC. It now lay in the middle of the South Downs Training Area.[49]

At Ditchling, 200 acres of rough hill grazing were taken over, and in Falmer it was said that 'the tenant's methods of farming have been greatly altered by the military taking about 200 acres of the holding'. The farmer's way of life was even more dramatically altered in Hartfield, where 15 acres of land were requisitioned for a military camp: 'The War Office bought the holding for a camp site. The previous owner now rents from the War Office his house and two acres of land … the whole of the land is a camp site. The tenant is working as a farm hand.' Another Army camp was created on farming land at Maresfield, where it was recorded that '10 acres went to the War Office'. More positive, perhaps, is the record from East Chiltington that pigs were kept on the swill from the local camp! Soldiers from the nearby camp are also recorded as harvest workers at Withyham.[50]

At Wiston, 300 acres taken by the army included Chanctonbury Ring and the area of downland around. More downland was requisitioned at Plumpton, and at Iford the military occupied 800 acres of hill land, 'some of which was arable'. A further 700–800 acres of another holding were requisitioned in Pyecombe. The military control of land in Findon was frustrating to the farmers: 'There are many acres of land on these farms that could be brought into cultivation if conditions allowed, i.e. army training.' At New Romney in Kent there was 'no land left suitable for ploughing' on one farm 'partly occupied by the War Department'. In South Heighton, 200 of the best acres were lost to the army.[51]

There is some reference in the records to the shortage of cottages as accommodation for farm workers, in particular when these were requisitioned for the army. In Angmering, on one holding it was recorded that a cottage that was occupied by the army was 'badly needed for the proper cultivation of the farm'. In Burpham a farm cottage was being lived in by the wife of a man who was now in the Army. Again, it was needed for farm workers. At West Bridgford (Nottinghamshire), two cottages had been condemned, but were 'being held in reserve for bombed out families'.[52]

In Hartfield, a house was requisitioned by the military, while at a large country house nearby the owner complained that 'the military have taken possession of nearly all the park land'. In Herefordshire, at Mansel Lacy, a private house was 'in occupation by Australian troops ... house and gardens are in occupation of Australian Forces so that no particulars can be gained'. At Tyneham in Dorset it is merely recorded for one holding that 'eight and a half acres has gone to the military' and this record therefore predates the mass requisitioning of the whole Tyneham area, and the evacuation of the civil population in the autumn of 1943, as a prelude to D-Day training in the area. The Tyneham farm records therefore depict the condition of the farms in the parish shortly before the military takeover, and would be a valuable addition to our understanding of the situation in the village just prior to its evacuation (Wright, 1996).[53]

Similarly, where the Census Returns for holdings taken over by the military survive in full for 4 June 1941, these provide a vivid snapshot of the production of the farm, and the loss to agriculture, just before the requisitioning orders were issued. Examples are in Stanmer, where the crop returns survive for a 650-acre farm taken over by the Army, and similarly at Westmeston. At Newtimber, most of a 1200-acre farm was requisitioned for the Army. A large dairy herd and a flock of Southdowns had been kept and these were now dispersed. The Census Return shows the exact numbers of the animals lost from this holding.[54]

The East Sussex and West Sussex map sheets record the extent of the military land forming the South Downs Training Area (Fig. 7.8). This land is shown by the words 'Military Area' or simply 'Military', with the boundaries generally emphasized with red ink. Some West Sussex sheets use the symbol of the upright broad arrow with the letters 'WD' to indicate military land. West Sussex LI NW has 'WD' and, for example, 'was 573' (the case number of the farm whose land has been requisitioned for the Army). The use of 'WD' is also found on sheet Pembrokeshire XL.1 as a pencilled annotation in fields around existing barracks.[55]

Other maps show 'Military Camps' or 'WD Camps' (East Sussex XXVII SE, Surrey XXXVII.8, Cornwall LXXII.13, Essex LXXII NW, Essex LXIII NE and Flintshire IV.10 are examples). Aerodromes are sometimes indicated by the use of that word on the map face. Buckinghamshire provides an example (Buckinghamshire XLVI NE), as do sheets Kent L.5, 9 and Berkshire

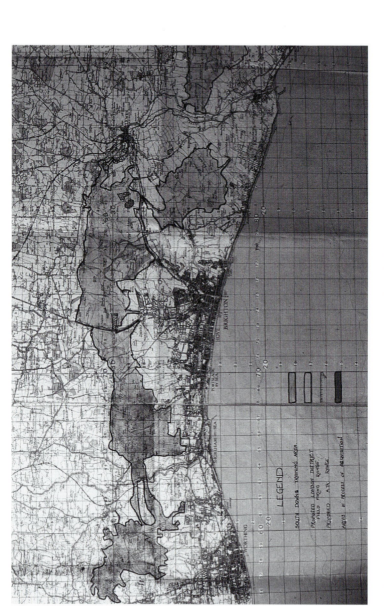

Fig. 7.8. The military training areas in the central South Downs (source: PRO MAF 48/394, is Crown copyright and is reproduced with the permission of the Controller of Her Majesty's Stationery Office).

XIV.5, 6, 9, 10, where the words 'Taken over by Air Ministry' are added as well. Similarly, 'Air Ministry' is used on sheets Staffordshire XXXVI.4, 8 and Huntingdonshire XII.15: in the latter case, 'Air Min. 1942' is added. Sheet West Sussex LII SW simply has 'RAF' on the map, as do Gloucestershire L.4, 8, Warwickshire XXXVIII.6, 10, Oxfordshire XXVIII.9, 13 and Oxfordshire XXVII.5, 9. On sheet Huntingdonshire XVII.4, an aerodrome is marked by the words 'Air Min.' and the descriptions 'Main Runway', 'Proposed Extension' and 'Proposed Overshoot'. Adjacent farm land is labelled 'Additional land for scheme'. East Sussex XXVIII SE shows 'WD Dump', and Devon CXVIII.16 has 'Ministry of Supply'. A Royal Ordnance Factory appears sited on former farmland in the parish of Ruddington (Nottinghamshire XLII and XLVI). Herefordshire XLI.4 and Huntingdonshire XIII.7 show the sites of prisoner-of-war camps.[56]

Farmers and their families in the armed forces and auxiliary services

The husband of a woman farmer in Hartfield is recorded as being in the Royal Air Force, as was the farmer husband of a woman left to manage their farm in Westmeston. There are many other examples of women being left to run the farms while their husbands were in the armed forces. At Burwash, the wife was trying to carry on, but 'not doing all that well', and the perception was similar in Plumpton where the farm was graded B− since 'husband in the forces. Wife is endeavouring to carry on the Farm but has insufficient knowledge of farming. Full productivity is affected owing to lack of capital and financial difficulties' (Short and Watkins, 1994: 290).[57]

Other records state merely that the farmer was in the armed forces, but do not say who was left to run the farm. In Llansaintffraid Glynceiriog (Denbighshire), two farms had farmers in the army (one as an officer), while in East Grinstead the farmer was a squadron-leader in the RAF. In Falmer, it was the farmer's son who had joined the Army. Many farmers served in the Home Guard, where their detailed and expert knowledge of the landscape was much valued. In Alfriston, the owner of a holding was the commanding officer of the local Home Guard unit. In Berwick the farmer was a captain in the Home Guard while in Buxted he was the 'platoon commander'. A Washington farm records two horses for the 'Home Guard patrol', while another farmer in that parish also belonged to the local unit. Yet another Home Guard farmer member was in Ditchling. Other auxiliary services on the Home Front are recorded. In Hartfield, the farmer was in the Royal Observer Corps, while in Maresfield he was an officer in the ARP. Finally, at Frensham (Surrey) a farmer was in the Auxiliary Fire Service.[58]

Enemy action against the farms

There are some references in the survey records to the damage caused by German bombs falling on the farms. In Compton, a 'bomb hole' is recorded on the farm, under 'All Other Crops' in the Census Return! The farmer was accounting for the loss of land caused by this unfilled crater. Four cows were killed by bombs in Maresfield, and in Plumpton cottages were destroyed by enemy action and a tractor damaged in an air raid. Of defences against German air attacks, only a searchlight on the rough grazing area of a farm is recorded at Angmering.[59]

The Plough-up Campaign as Seen in the Records of the National Farm Survey 1941–1943

One of the most important yet sensitive aspects of the CWAEC operation was the supervision of the plough-up campaign across England and Wales. Indeed, this campaign was at the very root of the NFS, since the surveyors had to ascertain the degree to which farms over 5 acres were capable of producing more foodcrops. Only secondarily was the survey designed as a 'Domesday Book'.[60] The process of implementation and the consequent results have been characterized in terms of their ecological impact, their economic impact, their social divisiveness, and their impact upon the rural landscape. Enormous expenditure of energy by the CWAECs, together with officials and scientists such as Sir George Stapledon, went into this campaign (Fig. 7.9).

The Primary Farm Return included within the NFS records does now allow a relatively complete appraisal of the spatial extent and pattern of the plough-up in the early years of the war, since each farm surveyed was required to include a record of the fields ploughed up in those years. It should be remembered, however, that the government was aiming to obtain a maximum tillage area by 1942/43, when more shipping would be required for the opening of the Second Front, and consequently that the data provided by the NFS does not allow us to see the position at its planned peak.[61]

The information contained in the Primary Return

The following information is generally available from Section F of the return.

Grass fields ploughed up: the harvests of 1940 and 1941

In a few cases the dates '1940' and '1941' have been altered to '1941' and '1942' respectively. In some cases, as in Warwickshire, another form has

Fig. 7.9. Sir George Stapledon with Rt Hon. Sir Reginald Dorman Smith, Minister of Agriculture (February 1939 to May 1940), addressing a wartime plough-up campaign meeting at the Somerset Farm Institute, Cannington (source: Reproduced by permission of Faber and Faber from R. Waller, *Prophet of the New Age* (1962) facing p. 240).

actually been locally printed over the Primary Return form for 1943. Fields
ploughed up for the 1940 harvest were presumably ploughed up during
1939, and for 1941 during 1940 – indeed there is some evidence in the
records that systematic ploughing up was going on even in 1938. However,
it is unclear whether some of the Surveyors misunderstood the instructions
and entered the wrong details under '1942' or '1943'. In all the discussion
below, figures given for the 1942 harvest must be treated with caution, and
even as an aberration within the records.

Field Ordnance Survey number and edition

The OS parcel numbers of the fields ploughed are entered here. Often
these are parts of fields, e.g. '115 Pt' or 'Pt. 115'. Between 18% and 38% of
the OS parcels returned as ploughed up gave an indication that a proportion of the field was ploughed up (Table 7.11). The date of the edition of
the relevant OS sheet should have been placed above the list of parcel
numbers. It was included on around 75% of occasions in the *National
Sample*, 95% of occasions in the *Midlands Sample*, and around 40% of
occasions in the *Sussex Sample*. Sometimes it was given simply as the year
of the edition, e.g. '1911', and on other occasions the sheet reference was
given in a variety of formats, e.g. 'Anglesey VII.13 (1920)', '47 NE 2nd ed
1905', 'VI – 2 1905', '42/2/14'. These were all the result of local practice. In
the parish of Hartfield, East Sussex, the edition was given simply as the
name of the parish. Sometimes the OS parcel numbers are bracketed
together, presumably if they form part of a single ploughing operation for
the same crop. The format of these references varies greatly, but some kind
of entry was given for between 39 and 96% of occasions when it should

Table 7.11. Number and percentage of occasions where it was indicated that part
of an OS parcel was ploughed up.

Sample	Year	Occasions when indicated that part of an OS parcel ploughed up	
		Number	Percentage
National Sample	1940	228	24
	1941	254	23
	1942	5	17
Sussex Sample	1940	269	38
	1941	277	32
	1942	7	29
Midlands Sample	1940	53	18
	1941	81	24
	1942	0	0

Table 7.12. Comparative figures for the presence of certain fields of data concerning the plough-up.

Database	Year	No. of entries for plough-up	Entries under 'Edition'		Entries under 'Parish'		Entries with acreage figure or area sown		Entries responding 'yes' to under CWAEC direction	
			No.	%	No.	%	No.	%	No.	%
National Sample	1940	960	746	78	951	99	634	66	842	88
	1941	1090	796	73	1079	99	643	59	1040	95
	1942	30	17	57	30	100	18	60	29	97
Sussex Sample	1940	711	307	43	699	98	299	42	497	70
	1941	864	341	39	858	99	341	39	574	66
	1942	24	2	8	23	96	2	8	21	88
Midlands Sample	1940	293	281	96	293	100	223	76	286	98
	1941	333	316	95	333	100	247	74	325	98
	1942	0	0	0	0	0	0	0	0	0

have been (Table 7.12). *The National Sample* returned around 75%, the *Sussex Sample* around 41% and the *Midlands Sample* 95%. When there was no entry this would often be for a whole parish, such as Gosforth (Cumberland), Isleham (Cambridgeshire) and Llantsaintffraid (Denbighshire) in the *National Sample* where surveyors in these areas had chosen not to follow the instructions in completing this section of the Primary Return.

Parish

The name of the parish in which the ploughed field lies is entered here. This was returned with a very high degree of consistency across all the samples, not dropping below 98%. Sometimes the acreage of the field is also added here, before the parish name.

Crops sown

The crops grown in the ploughed field are entered here, e.g. 'wheat', 'oats' or 'rape and mustard'. The phrase 'approved crop' is also used. Sometimes not a crop as such but alternative details such as 're-seeded' or 'fallow' are given. A very wide variety of types of crop and many different combinations were given so that simple quantitative summaries are not easy to give, while some generic names were also used, such as 'cereals', 'green crops' and 'vegetables'.

The occurrences of a variety of the main types of crops specified are given in Tables 7.13 to 7.15. In the *National Sample*, information was returned under this section on between 95.5 and 99.4% of all occasions. In the *Sussex Sample*, information was returned under this section on 98–98.3% of all occasions; in the *Midlands Sample*, information was returned under this section on every occasion.

Sometimes the acreage of the field is also entered here, although there was no stipulation to do so. In addition West Sussex records frequently have extra information in this column, such as 'Certified to M.A.F. 12 acres'. Sometimes this is bracketed to show that the information relates to several fields (parcel numbers). An acreage figure might be given for crops where more than one was specified, or the proportion of the land ploughed up given over to various crops might be given, most often as $\frac{1}{2}$ kale, $\frac{1}{2}$ potatoes', for example. There were no clear conventions, however, and any figures given might not be very reliable for further analysis.

The frequency with which acreage figures were given varied greatly, with around 60% of returns giving an acreage figure in the *National Sample*, around 40% in the *Sussex Sample* and around 74% in the *Midlands Sample*. It could be a simple number or to three decimal places, which probably indicates that it was derived from the OS parcel acreage figures given on the 1:2500 maps.

Table 7.13. Number of occurrences and percentage of all occurrences of various crops planted on ploughed-up land from the National Sample.

Crop(s)	1940 No.	1940 %	1941 No.	1941 %	1942 No.	1942 %	1940 and 1941 No.	1940 and 1941 %
No information given	43	4.5	7	0.6	1	3.3	50	2.4
Some information given	917	95.5	1083	99.4	29	96.7	2000	97.6
Approved crop	21	2.2	44	4.0	1	3.3	65	3.2
Oats only	535	55.7	596	54.7	5	16.7	1131	55.2
All occurrences of oats	623	64.9	711	65.2	5	16.7	1334	65.1
Oats and wheat	20	2.1	4	0.4	**0**	**0.0**	24	1.2
Oats and barley	4	0.4	5	0.5	**0**	**0.0**	9	0.4
Spring oats	2	0.2	10	0.9	**0**	**0.0**	12	0.6
Winter oats	**0**	**0.0**	1	0.1	**0**	**0.0**	1	0.0
Oats and other combinations	64	6.7	106	9.7	**0**	**0.0**	157	7.7
Wheat only	83	8.6	71	6.5	18	60.0	154	7.5
All occurrences of wheat	126	13.1	95	8.7	18	60.0	221	10.8
Wheat and barley	2	0.2	1	0.1	**0**	**0.0**	3	0.1
Wheat and beans	3	0.3	1	0.1	**0**	**0.0**	4	0.2
Wheat and potatoes	3	0.3	3	0.3	**0**	**0.0**	6	0.3
Spring wheat	3	0.3	**0**	**0.0**	**0**	**0.0**	3	0.1
Winter wheat	**0**	**0.0**	**0**	**0.0**	**0**	**0.0**	**0**	**0.0**
Wheat and other combinations	32	3.3	19	1.7	**0**	**0.0**	51	2.5
Barley only	12	1.3	19	1.7	**0**	**0.0**	31	1.5
All occurrences of barley	27	2.8	33	3.0	**0**	**0.0**	60	2.9
Barley and other crops	21	2.2	14	1.3	**0**	**0.0**	35	1.7
Oats, wheat and barley	2	0.2	5	0.5	**0**	**0.0**	7	0.3
All occurrences of corn	17	1.8	26	2.4	**0**	**0.0**	43	2.1
Spring corn	2	0.2	4	0.4	**0**	**0.0**	6	0.3
Dredgecorn	12	1.3	7	0.6	**0**	**0.0**	19	0.9
Cereals only	27	2.8	17	1.6	**0**	**0.0**	44	2.1
All occurrences of cereals	28	2.9	21	1.9	**0**	**0.0**	49	2.4
Cereals and other crops	1	0.1	4	0.4	**0**	**0.0**	5	0.2
All occurrences of beans	7	0.7	22	2.0	1	3.3	29	1.4
Potatoes only	33	3.4	39	3.6	1	3.3	72	3.5
All occurrences of potatoes	65	6.8	89	8.2	1	3.3	154	7.5
Potatoes and other crops	32	3.3	50	4.6	**0**	**0.0**	82	4.0
Rape only	8	0.8	12	1.1	**0**	**0.0**	20	1.0
All occurrences of rape	22	2.3	35	3.2	**0**	**0.0**	57	2.8
Rape and other crops	14	1.5	23	2.1	**0**	**0.0**	37	1.8
Roots only	14	1.5	28	2.6	2	6.7	42	2.0
All occurrences of roots	33	3.4	56	5.1	2	6.7	89	4.3
Roots and other crops	19	2.0	28	2.6	**0**	**0.0**	47	2.3
All occurrences of turnips	5	0.5	4	0.4	**0**	**0.0**	9	0.4
All occurrences of kale	15	1.6	38	3.5	**0**	**0.0**	53	2.6
Green crops/greenstuff	4	0.4	5	0.5	**0**	**0.0**	9	0.4
All occurrences of linseed	3	0.3	**0**	**0.0**	**0**	**0.0**	3	0.1
Re-seed(ed)	8	0.8	18	1.7	**0**	**0.0**	26	1.3
Fallow	5	0.5	8	0.7	**0**	**0.0**	13	0.6

Bold indicates no plough-up.

Table 7.14. Number of occurrences and percentage of all occurrences of various crops planted on ploughed-up land from the Sussex Sample.

Crop(s)	1940 No.	1940 %	1941 No.	1941 %	1942 No.	1942 %	1940 and 1941 No.	1940 and 1941 %
No information given	14	2.0	15	1.7	**0**	**0.0**	29	1.8
Some information given	697	98.0	849	98.3	24	100.0	1546	98.2
Approved crop	40	5.6	33	3.8	1	4.2	73	4.6
Oats only	219	30.8	278	32.2	6	25.0	497	31.6
All occurrences of oats	319	44.9	404	46.8	9	37.5	723	45.9
Oats and wheat	25	3.5	11	1.3	**0**	**0.0**	36	2.3
Oats and barley	4	0.6	21	2.4	**0**	**0.0**	25	1.6
Spring oats	19	2.7	9	1.0	**0**	**0.0**	28	1.8
Winter oats	2	0.3	**0**	**0.0**	**0**	**0.0**	2	0.1
Oats and other combinations	50	7.0	85	9.8	3	12.5	135	8.6
Wheat only	132	18.6	109	12.6	1	4.2	241	15.3
All occurrences of wheat	186	26.2	139	16.1	1	4.2	325	20.6
Wheat and barley	9	1.3	1	0.1	**0**	**0.0**	10	0.6
Wheat and beans	2	0.3	1	0.1	**0**	**0.0**	3	0.2
Wheat and potatoes	**0**	**0.0**	**0**	**0.0**	**0**	**0.0**	**0**	**0.0**
Spring wheat	**0**	**0.0**	1	0.1	**0**	**0.0**	1	0.1
Winter wheat	**0**	**0.0**	**0**	**0.0**	**0**	**0.0**	**0**	**0.0**
Wheat and other combinations	38	5.3	27	3.1	**0**	**0.0**	65	4.1
Barley only	22	3.1	25	2.9	1	4.2	47	3.0
All occurrences of barley	46	6.5	78	9.0	1	4.2	124	7.9
Barley and other crops	24	3.4	53	6.1	**0**	**0.0**	77	4.9
Oats, wheat and barley	**0**	**0.0**	5	0.6	**0**	**0.0**	5	0.3
All occurrences of corn	4	0.6	39	4.5	**0**	**0.0**	43	2.7
Spring corn	1	0.1	2	0.2	**0**	**0.0**	3	0.2
Dredgecorn	**0**	**0.0**	11	1.3	**0**	**0.0**	11	0.7
Cereals only	**0**	**0.0**	**0**	**0.0**	**0**	**0.0**	**0**	**0.0**
All occurrences of cereals	**0**	**0.0**	**0**	**0.0**	**0**	**0.0**	**0**	**0.0**
Cereals and other crops	**0**	**0.0**	**0**	**0.0**	**0**	**0.0**	**0**	**0.0**
All occurrences of beans	37	5.2	22	2.5	**0**	**0.0**	59	3.7
Potatoes only	16	2.3	5	0.6	**0**	**0.0**	21	1.3
All occurrences of potatoes	37	5.2	21	2.4	**0**	**0.0**	58	3.7
Potatoes and other crops	21	3.0	16	1.9	**0**	**0.0**	37	2.3
Rape only	3	0.4	1	0.1	**0**	**0.0**	4	0.3
All occurrences of rape	7	1.0	4	0.5	**0**	**0.0**	11	0.7
Rape and other crops	4	0.6	3	0.3	**0**	**0.0**	7	0.4
Roots only	14	2.0	21	2.4	3	12.5	35	2.2
All occurrences of roots	43	6.0	44	5.1	3	12.5	87	5.5
Roots and other crops	29	4.1	23	2.7	**0**	**0.0**	52	3.3
All occurrences of turnips	2	0.3	4	0.5	**0**	**0.0**	6	0.4
All occurrences of kale	38	5.3	54	6.3	**0**	**0.0**	92	5.8
Green crops/greenstuff	8	1.1	3	0.3	**0**	**0.0**	11	0.7
All occurrences of linseed	8	1.1	18	2.1	**0**	**0.0**	26	1.7
Re-seed(ed)	15	2.1	11	1.3	4	16.7	26	1.7
Fallow	16	2.3	21	2.4	**0**	**0.0**	37	2.3

Bold indicates no plough-up.

Table 7.15. Number of occurrences and percentage of all occurrences of various crops planted on ploughed-up land from the Midlands Sample.

Crop(s)	1940 No.	1940 %	1941 No.	1941 %	1942 No.	1942 %	1940 and 1941 No.	1940 and 1941 %
No information given	**0**	**0.0**	**0**	**0.0**	**0**	**0.0**	**0**	**0.0**
Some information given	293	100.0	333	100.0	**0**	**0.0**	626	100.0
Approved crop	5	1.7	5	1.5	**0**	**0.0**	10	1.6
Oats only	146	49.8	157	47.1	**0**	**0.0**	303	48.4
All occurrences of oats	192	65.5	204	61.3	**0**	**0.0**	396	63.3
Oats and wheat	2	0.7	1	0.3	**0**	**0.0**	3	0.5
Oats and barley	1	0.3	2	0.6	**0**	**0.0**	3	0.5
Spring oats	**0**	**0.0**	**0**	**0.0**	**0**	**0.0**	**0**	**0.0**
Winter oats	1	0.3	**0**	**0.0**	**0**	**0.0**	1	0.2
Oats and other combinations	42	14.3	44	13.2	**0**	**0.0**	86	13.7
Wheat only	25	8.5	13	3.9	**0**	**0.0**	38	6.1
All occurrences of wheat	42	14.3	22	6.6	**0**	**0.0**	64	10.2
Wheat and barley	**0**	**0.0**	1	0.3	**0**	**0.0**	1	0.2
Wheat and beans	**0**	**0.0**	**0**	**0.0**	**0**	**0.0**	**0**	**0.0**
Wheat and potatoes	**0**	**0.0**	**0**	**0.0**	**0**	**0.0**	**0**	**0.0**
Spring wheat	**0**	**0.0**	1	0.3	**0**	**0.0**	1	0.2
Winter wheat	**0**	**0.0**	**0**	**0.0**	**0**	**0.0**	**0**	**0.0**
Wheat and other combinations	17	5.8	7	2.1	**0**	**0.0**	24	3.8
Barley only	8	2.7	3	0.9	**0**	**0.0**	11	1.8
All occurrences of barley	9	3.1	8	2.4	**0**	**0.0**	17	2.7
Barley and other crops	1	0.3	5	1.5	**0**	**0.0**	6	1.0
Oats, wheat and barley	**0**	**0.0**	**0**	**0.0**	**0**	**0.0**	**0**	**0.0**
All occurrences of corn	**0**	**0.0**	**0**	**0.0**	**0**	**0.0**	**0**	**0.0**
Spring corn	**0**	**0.0**	**0**	**0.0**	**0**	**0.0**	**0**	**0.0**
Dredgecorn	**0**	**0.0**	**0**	**0.0**	**0**	**0.0**	**0**	**0.0**
Cereals only	1	0.3	26	7.8	**0**	**0.0**	27	4.3
All occurrences of cereals	1	0.3	27	8.1	**0**	**0.0**	28	4.5
Cereals and other crops	**0**	**0.0**	1	0.3	**0**	**0.0**	1	0.2
All occurrences of beans	1	0.3	3	0.9	**0**	**0.0**	4	0.6
Potatoes only	14	4.8	12	3.6	**0**	**0.0**	26	4.2
All occurrences of potatoes	27	9.2	49	14.7	**0**	**0.0**	76	12.1
Potatoes and other crops	13	4.4	37	11.1	**0**	**0.0**	50	8.0
Rape only	2	0.7	5	1.5	**0**	**0.0**	7	1.1
All occurrences of rape	6	2.0	15	4.5	**0**	**0.0**	21	3.4
Rape and other crops	4	1.4	10	3.0	**0**	**0.0**	14	2.2
Roots only	1	0.3	1	0.3	**0**	**0.0**	2	0.3
All occurrences of roots	10	3.4	22	6.6	**0**	**0.0**	32	5.1
Roots and other crops	9	3.1	21	6.3	**0**	**0.0**	30	4.8
All occurrences of turnips	3	1.0	0	0.0	**0**	**0.0**	3	0.5
All occurrences of kale	13	4.4	13	3.9	**0**	**0.0**	26	4.2
Green crops/greenstuff	3	1.0	16	4.8	**0**	**0.0**	19	3.0
All occurrences of linseed	**0**	**0.0**	**0**	**0.0**	**0**	**0.0**	**0**	**0.0**
Re-seed(ed)	6	2.0	15	4.5	**0**	**0.0**	21	3.4
Fallow	**0**	**0.0**	**0**	**0.0**	**0**	**0.0**	**0**	**0.0**

Bold indicates no plough-up.

On some occasions this figure would be accompanied by 'Pt.', indicating that part of the OS parcel had been ploughed up, and on other occasions there might be another acreage figure indicating what proportion of the parcel had been ploughed up. Further indication in this section might have been given of what proportion of the area ploughed up was sown with different crops, if more than one had been indicated in the column for crops, along with information as to why it may have failed, although this only occurred very occasionally. These were all the result of local practice and so do not provide a firm basis for any meaningful comparisons. However, in a parish or region where fuller details were given there is great potential for mapping the extent of the plough-up in 1940 and 1941 (or even 1942) and calculating its extent and proportion of agricultural land available. In the context of other information a full picture of this important change in land use can be developed.

In one parish, Shaugh Prior in Devon, there was a figure of money consistently noted down in the following format 'PP £2 4.5 acres [part]' or 'TP £2 2.25 acres'. This must have been the subsidy paid for ploughing up the land recorded in the entry, but the practice of recording such information was not widespread.

Under W.A.E.C.'s direction? Yes/No

Almost invariably the answer to this question is 'Yes', indicated by a cross in the appropriate box. The response 'Yes' was given on around 90% of occasions in the *National Sample*, around 68% in the *Sussex Sample*, and 98% in the *Midlands Sample*. However, if the entry in 'Crops Sown' (see above) provides information on the crop being 'Certified to M.A.F.', then there is usually no Yes/No entry (this occurred mostly in East and West Sussex). Sometimes the field was also left blank.

Some or all of the details required under Section F of the Primary Return dealing with the plough-up might be omitted, and on many occasions a figure might be given for the acreage either of the field or of the crops which were sown. The *National Sample* seems to provide a level of provision of the required data (plus the acreage figure) of between 60 and 90%; the *Sussex Sample* of between 40 and 60%, and the *Midlands Sample* between 75 and 95%.

The national and regional extent of the plough-up campaign

In the *National Sample,* 39% of all holdings returned some information concerning the plough-up for the 1940 harvest, 44% for 1941, and 1% for 1942 (although this was not required information), and for holdings returning information for both 1940 and 1941 it was 31%. In the *Sussex Sample,* 29%

Table 7.16. Number and percentage of holdings giving a return of some land ploughed up 1940, 1941 and 1942.

		Holdings giving a return of some land ploughed up	
Sample	Year(s)	Number	Percentage
National Sample	1940	517	39
	1941	580	44
	1942	14	1
	1940 and 1941	412	31
	1941 and 1942	13	1
	1940, 1941 and 1942	10	1
Sussex Sample	1940	340	29
	1941	374	32
	1942	12	1
	1940 and 1941	217	18
	1941 and 1942	2	0
	1940, 1941 and 1942	0	0
Midlands Sample	1940	173	36
	1941	191	40
	1942	0	0
	1940 and 1941	128	27
	1941 and 1942	0	0
	1940, 1941 and 1942	0	0

of all holdings returned some information concerning the plough-up for the 1940 harvest, 32% for 1941, and 1% for 1942, and for both 1940 and 1941 it was 18%. In the *Midlands Sample*, 36% of all holdings returned some information concerning the plough-up for the 1940 harvest, 40% for 1941, and for both 1940 and 1941 it was 27% (Table 7.16).

These figures are fairly consistent across all three samples. This is, perhaps, to be expected as CWAECs had been instructed to plough up a proportion of the agricultural land under grass in their area and were assessed as to their performance. The plough-up was one of the more closely monitored aspects of agricultural activity during the war, as it reflected the drive to increase agricultural production and involved the payment of subsidies for any ploughing undertaken, although they did not always make crop production economical, depending on the area and the type of land ploughed up.[62]

It is more surprising at first sight that a higher proportion of holdings did not plough up land in East and West Sussex, but this reflects the fact that Sussex was already relatively intensively cultivated despite the period of agricultural depression leading up to the war, perhaps due to its proximity to London, and had less slack to take up than other counties in terms of the plough-up, that there were also areas of heavy or difficult soils, and perhaps because of the amounts of land appropriated by the military. A recent study of the Sussex downland indicates figures of 5% and 4.5% of land being ploughed up in 1940 and 1941, respectively – some way below the national target. The large area of Sussex downland taken for military use was reinstated in 1946/47 when armour-plated cabs for the Fordson tractors were issued.[63]

There was quite a high degree of regional variation in the proportion of holdings which returned some ploughed-up land for the 1940 and 1941 harvests. Amongst the Advisory Centre Provinces, the proportion of holdings returning some ploughed-up land for the 1940 harvest varied from 13% in Yorkshire to 64% in Wales (and 68% in North Wales). Overall, the Northern, South Western and Wales (both North and South) Advisory Centre Provinces returned significantly higher proportions than for the *National Sample* as a whole, while the Eastern, Southern, West Midland, Western and Yorkshire Advisory Centre Provinces returned significantly lower proportions than for the *National Sample* as a whole. The East Midland, North Western and South Eastern Advisory Centre Provinces returned around the same proportions as the *National Sample*. England as a whole had less than half the proportion of holdings returning some ploughed-up land as Wales (Table 7.17).

The proportions for the 1941 harvest were largely similar, with most of the Advisory Centre Provinces increasing the proportion of their holdings returning some ploughed-up land by between 1% and 13%. For the *National Sample* as a whole, 5% more of the holdings ploughed up land for the 1941 harvest than for 1940, and this pattern was followed for both England and Wales as a whole, although once again the figure for England was only half that for Wales. Most of the Advisory Centre Provinces followed this pattern of an increase between 1940 and 1941 except the North Western, Northern and South Eastern. The Western and Yorkshire Advisory Centre Provinces experienced a greater than average increase in the proportion of holdings returning ploughed-up land between 1940 and 1941 (Table 7.17).

The proportion of holdings that ploughed up land for both the 1940 and 1941 harvests varied from between 6% for Yorkshire and 58% for Wales (with South Wales being 62%). The Eastern, South Eastern, Southern, West Midland, Western and Yorkshire Advisory Centre Provinces returned lower proportions than the *National Sample* as a whole, and the Northern and South Western Advisory Centre Provinces returned significantly higher proportions. The East Midland and North Western Advisory Centre Provinces were about the same as the *National Sample*. Here the disparity between England and Wales was even more marked, with England returning only 22% of holdings ploughing up in both years, whereas Wales returned 58%. Although Wales comprised only 24% of all the holdings in the *National Sample*, it returned 39% of the holdings with land ploughed up for the 1940 harvest, 38% for the 1941 harvest and 45% for both. England comprised 76% of the *National Sample*, but returned only 61% for the 1940 harvest, 62% for the 1941 harvest and 55% for both. The extent of the plough-up in Wales, where all counties exceeded their plough-up quotas, may be accounted for

Table 7.17. Number and percentage of holdings returning some land ploughed up, by Advisory Centre Province, 1940 and 1941.

(a) England.

Advisory Centre Province	Year(s)	No. of holdings giving a return for plough-up	% of all holdings
East Midland	1940	53	43
	1941	65	53
	1940 and 1941	42	34
Eastern	1940	32	18
	1941	47	27
	1940 and 1941	17	10
North Western	1940	12	39
	1941	10	32
	1940 and 1941	10	32
Northern	1940	50	56
	1941	50	56
	1940 and 1941	44	49
South Eastern	1940	43	32
	1941	26	27
	1940 and 1941	27	20
South Western	1940	40	63
	1941	43	67
	1940 and 1941	34	53
Southern	1940	34	25
	1941	35	26
	1940 and 1941	20	15
West Midland	1940	15	22
	1941	17	25
	1940 and 1941	11	16
Western	1940	21	21
	1941	30	31
	1940 and 1941	15	15
Yorkshire	1940	13	13
	1941	25	26
	1940 and 1941	6	6
England	1940	313	31
	1941	358	35
	1940 and 1941	226	22

Table 7.17. (cont.)

(b) Wales.

Advisory Centre Province	Year(s)	No. of holdings giving a return for plough-up	% of all holdings
Wales	1940	204	64
	1941	222	69
	1940 and 1941	186	58
North Wales	1940	112	61
	1941	125	68
	1940 and 1941	102	55
South Wales	1940	92	68
	1941	97	71
	1940 and 1941	84	62
National Sample	1940	517	39
	1941	580	44
	1940 and 1941	412	31

by some of the large experimental schemes run by the Ministry, such as the Dolfor scheme in Montgomeryshire where previously unploughed upland areas on the boundaries of viable cultivation were ploughed up using newly available American heavy tractors. It is hard to imagine, however, that such schemes fully account for the disparity between England and Wales revealed by the above figures (Ministry of Information, 1945b).[64]

Although the number of holdings ploughing up land and the number of OS parcels returned as ploughed up does not indicate how much land in terms of acres was brought back into active production – or ploughed for the first time – it does give an idea of the extent of the plough-up and its impact upon the general farming population.

The plough-up at the local and holding level

It is a rather more difficult task to judge the amount of land ploughed up on individual holdings, as an acreage figure on the OS parcels ploughed up was not stipulated for the Primary Return. This is a strange and unfortunate omission, as it would be fundamental in understanding and monitoring the effectiveness of the plough-up campaign. However, the information was certainly recorded elsewhere in the files kept by the CWAECs, as each field had to be scheduled for ploughing up; the CWAEC may well have overseen the actual ploughing and a subsidy would be paid to the holder. Even by 1945, the Ministry of Information was able to give a national total of land ploughed up since 1939 of 6.5 million acres, so the data were available to them at that stage.[65] Nothing concerning the plough-up appears in the 1946 Summary Report of the NFS. It is unfortunate that such an important figure for understanding the changes brought to agriculture by the war was omitted from the NFS when it could so easily have been included, and that we have been left with piecemeal information that is unclear, given by surveyors of the CWAECs, who obviously thought that it should have been included in the Survey.

It is in fact likely that Section F of the Primary Return was completed from information that the CWAECs already had on file. This might account for the frequent presence of acreage figures to three decimal points, drawn from the 1:2500 map sheets, or which would have been transcribed on to the 1:10,560 sheets. Although between 39% and 76% of all OS parcels returned some kind of acreage figure under Section F, it is impossible: (i) to confirm whether or not this was the acreage actually ploughed up; (ii) to ascertain whether the acreage given was that of the OS parcel (this is most likely if it was the figure to three decimal places) or the amount of land ploughed up; or (iii) to determine whether all of that parcel might have been ploughed up or simply part of it, although our present survey indicated that on between 17% and 38% of occasions it was a part of the OS parcel that was ploughed up. But acreage figures were not always given on these occasions.

Local research certainly can produce a picture of the state of the plough-up campaign using Section F material. The strip parish of Ditchling, in Sussex, for example, straddles the chalk of the South Downs and the lower-lying clays to the north. The fields that were ploughed for the harvests of 1940 and 1941 are listed by OS parcel number, and the crops sown are indicated. A preliminary analysis suggests that most of the plough-up was concentrated around the village itself and on the lower-lying clays, rather than on the chalk. This is of interest if only because of the locally-held view that the Downs were transformed quite radically to an arable monoculture during the Second World War, whereas the picture emerging from the NFS at least cautions against too ready an acceptance of the assumed spatial incidence of the plough-up and clearly it was not all concentrated on the downland by any means (Fig. 7.10).

This picture can be amplified by reference to detailed work at rural district and county levels which was published in the individual county reports of the Land Utilisation Survey. The Report on Oxfordshire, for example, shows the results of the plough-up instructions given in the first year of the war (Fig. 7.11). The map demonstrates that the plough-up was undertaken all over the county, irrespective of soil type, although later in the war it was the grassland that yielded the bulk of the newly ploughed-up acreage (Marshall, 1943). Thus more widespread was the pattern as revealed in the Report on Buckinghamshire, which showed pre-war arable fields and plough-ups in 1939/40 and 1940/41 for Wing and Amersham Rural Districts, demonstrating a greater concentration in those areas that had formerly been

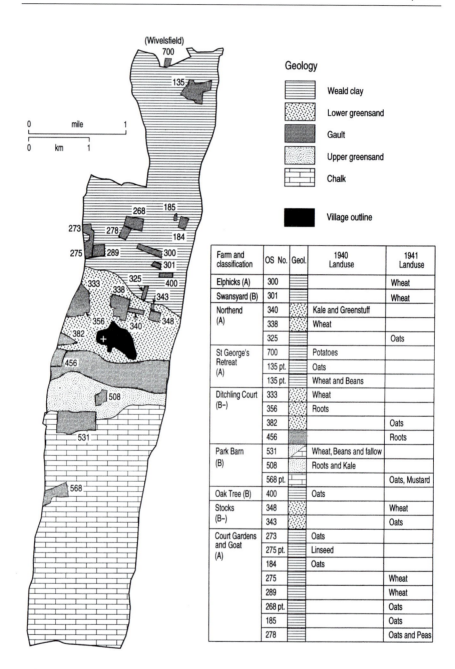

Fig. 7.10. The pattern of plough-up in Ditchling, East Sussex, 1940/41.

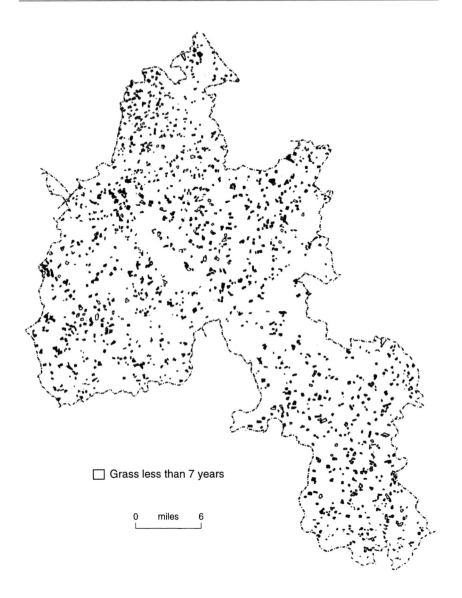

Fig. 7.11. The land ploughed in Oxfordshire 1939/40 (source: adapted from Marshall *et al.*, 1943).

grassland – the heavy clays of the county (Fryer, 1942) (Fig. 7.12). Similar again was the situation in East Anglia, where a group of farms in central Norfolk, two-thirds arable before the war, increased their arable area by just 4%, whereas a group on the heavy clays of south Essex increased their

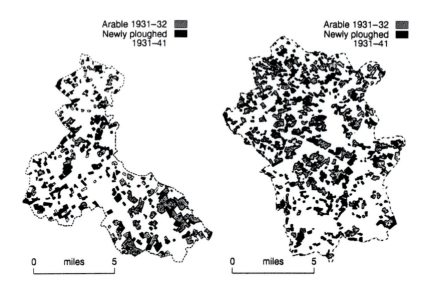

Fig. 7.12. Wing and Amersham Rural Districts, Buckinghamshire: the spatial incidence of the plough-up campaigns 1939/40 and 1940/41 (source: adapted from Fryer, 1942).

arable area from 26% to 54% (Menzies-Kitchen and Chapman, 1946: 37–85). The situation was summarized by Stamp in a paper read to the Royal Geographical Society in June 1946 (Stamp, 1947).

The crops sown on the ploughed-up land

Tables 7.13–7.15 detail the crop varieties sown, and the main categories of crop and the problems of information being available by generic type are noted above. There were a few crops that were by far the most significant. Because profit levels for oats and barley were higher than for wheat by the spring of 1940, the national picture was for nearly 50% of the extra 2 million acres ploughed to be sown with oats. At this stage market forces still dominated decisions about which crops to plant, and the Primary Returns therefore indicate that oats were both the single most significant crop and also important in various combinations with other crops. For both 1940 and 1941, in the *National Sample* around 55% of all the fields recorded as being ploughed up were sown exclusively with oats. A further 10% were sown with some kind of combination of crops which included oats (such as oats and wheat; oats and barley; oats, wheat and barley; oats and potatoes). In the *Sussex Sample* the figure was lower, at 32% and 14% respectively, and in the *Midlands Sample* they were 48% and 15%. Wheat was the next most important crop: in the *National Sample* around 8% of all the fields recorded as being ploughed up were sown exclusively with wheat. A further 3% were sown with some kind of combination of crops which included wheat; in the *Sussex Sample* the figures were 15% and 5%; in the *Midlands Sample* the figures were 6% and 4%. In all three samples the percentage of fields ploughed up and sown with oats, or some combinations including oats, increased between 1940 and 1941, and the figures for fields sown with wheat all decreased, even though by the summer of 1941 the CWAECs were better organized to produce more import-saving crops such as wheat, potatoes and sugar beet. It is worth noting that in the few instances of a crop being returned for 1942, with the rhetoric of propaganda at full blast to produce such crops, 60% of the returns mention wheat only. Within the Primary Returns, other crops of some significance were potatoes (which developed from 4% to 12% between the 2 years), and also roots (4–6%), approved crops (3–5%), barley (3–5%), corn (2–3%), re-seed (2–3%), beans (1–4%) and green crops/greenstuff/green vegetables (1–3%). All others were below 2%. Therefore, although a very wide range of types of crops and combinations were sown on land ploughed up, a few crops accounted for around 80% of the fields being newly sown. Some fodder crops, such as mangolds, roots, kale, etc., were grown predominantly in combinations with other crops, whereas crops such as cereals were grown mostly on their own.

The plough-up and the involvement of the CWAECs

The plough-up campaign has been characterized by Martin (1992: 184–185) as follows:

> Paradoxically … both too ambitious and too conservative at one and the same time. Too ambitious in that it attempted to convert designated areas of marginal land to tillage irrespective of the financial cost involved. Too conservative in that it stuck rigidly to an institutional framework of control in which the conversion of pasture to tillage was undertaken without sufficient consideration being given to the individual circumstances of the farmer.

These individual circumstances could mean that farmers and CWAECs were not always in harmonious relationship, and that parts of or even whole farms could be taken over by the local committee to facilitate the plough-up. The Primary Return will reflect this by having the fact stated under 'Farmer's Name' or occasionally in the tenure section: in East Sussex 18 holdings in the *Sussex Sample* were so designated (2.2% of the sample) and in West Sussex six holdings (1.5%). Most information on the farming of the land by the CWAEC, or the fact that the land may have to be taken over by the committee, is contained in the 'General Comments' section of the form. Although CWAECs had been instructed to include records of the holdings

that were under their own management, this does not always appear to
have been done, and where the ultimate sanction of dispossession has been
applied, a record of the farm under a new tenant has not always been
made. The notorious George Walden case at Itchen Stoke, Hampshire,
where Walden died in July 1940 resisting dispossession, is poorly served by
the NFS since the farm is missing from the Survey records and there is no
indication of anything unusual on the relevant map sheet, although it is
clear from the minutes of the Hampshire CWAEC that a new tenant had
been in occupation from August 1940.

On the Primary Return, other than in Section F, fields ordered to be
ploughed up are sometimes detailed, as at Buxted, Alciston and West Dean
in Sussex. Reseeding work ordered by the CWAEC at Buxted had failed, and
it was recommended that future work on this holding be supervised by a
member of the committee. In Burwash the 'only hope' was for the CWAEC
to plough an entire farm and at Denny Lodge (Hampshire) it was similarly
recorded that the holding would have to be cultivated by the Hampshire
WAEC. At East Grinstead derelict land was being taken over and 'land girls
are being used to advantage'. In the parish of Friston in East Sussex, and
spreading over adjoining West Dean and Folkington, the WAEC ran one of
its largest projects on what was known as the 'Friston land' – 1600 acres of
gorse-covered downland that had been worth as little as 5s per acre. Some
land was also taken over from the Forestry Commission and, using a labour
force of some 57 men, largely unskilled, the area was converted into flourishing corn fields, including 718 acres of wheat, 160 acres of barley and 204
acres of oats.[66]

Other evidence for CWAEC control to ensure plough-up compliance can
be found in the Census Return, where page 1 or page 3 may be annotated
to show the degree of control by a CWAEC, or the address may be altered
to show that of the CWAEC executive officer. The Horticultural Return
(C51/SSY) may similarly give this information. The maps also indicate land
under CWAEC control (Fig. 6.2) by some abbreviated reference written on
the sheet, or by a rather more informative comment within the margins.
Thus Middlesex XIX NE has 'Taken over by WAEC 1942'.

Other questions

Many other questions might now be profitably investigated concerning the
types of crops being newly planted. For example, what area of those crops
was sown? What were the most significant cropping changes between years?
How did these vary spatially between regions or Advisory Centre Provinces?
To what other purposes might such information on the plough-up be put?
One immediate answer must be that it provides important ecological information for those scholars interested in reconstructing the immediate past of

sensitive landscape areas. The ploughing was not always successful, and thus one finds from West Firle in Sussex, 'crop of oats on ploughed up grassland completely ruined by wireworm and leather jackets', or it was somewhat hopeful as with the 'strips of land among the rocks' on an upland farm in Uwchygarreg (Montgomeryshire).[67] To be able to trace the fields ploughed at this time is an enormous help in environmental reconstruction and interpretation work, although the limitations of the data explored above must be recognized (Stamp, 1947; Martin, 1992: 168–185).

Conclusion

This chapter has demonstrated the enormous research potential hidden within the many thousands of documents and maps that form the NFS. The great value of this source of data lies in its comprehensive coverage of almost all farms in England and Wales and the depth of information available for each holding. We have only touched on the implications and value of the NFS for the themes discussed in this chapter. And, of course, there are many other themes for which it has immediate and direct relevance including environmental change, family history, farm building history, and changes in farm enterprises and farm businesses. The NFS is clearly of enormous value when used on its own; its value is increased dramatically when it is combined with other sources, such as estate records. The depth and coverage of the NFS mean that it is as important a source for studying the specific history of an individual farm or parish as it is an essential benchmark for measuring broad developments in 20th century agriculture and rural society.

Notes

1   Public Record Office (PRO) MAF 32/772 (11).
2   PRO MAF 32/751 (47).
3   PRO MAF HF 325/236 (3).
4   PRO MAF HF 325/226 (5).
5   PRO MAF HF 325/226 (3).
6   PRO MAF 32/310 (67).
7   PRO MAF 32/1311 (21).
8   PRO MAF 32/1002 (1).
9   PRO MAF 32/1194 (36).
10 Arlesey, PRO MAF 32/751 (47).
11 Ardington, PRO MAF 32/865 (144).
12 Ruddington, PRO MAF 32/376 (161).
13 Edale, PRO MAF 32/321 (144); Beeley, MAF 32/319 (143); Stoney Middleton, MAF 32/340 (107); and Hope Woodlands, MAF 32/327 (148).
14 PRO MAF 32/370 (32).
15 Mansel Lacy, PRO MAF 32/17 (227); Wormsley, MAF 32/27 (235); Yazor, MAF 32/27 (236); Moccas, MAF 32/018 (226); Mansel Gamage, MAF 32/17 (225); and Bishopstone, MAF

32/2 (213). Ralph T. Hinckes was the author of *The Farmer's Outlook. A review of Home and Overseas Agriculture, 1880–1913*, Jarrold, London.
16 PRO MAF 32/751 (47).
17 Shepreth, PRO MAF 32/813 (23); Stanwell, MAF 32/954 (43); Stapleford and Beeston, MAF 32/350 (128); Little Thetford, MAF 32/772 (11).
18 Derwent, PRO MAF 32/319 (143); Hope Woodlands, MAF 32/327 (148).
19 PRO MAF 32/350 (128).
20 Denny Lodge, PRO MAF 32/974 (181); Machynlleth, MAF 32/1354 (38); East Grinstead, MAF 32/1006 (22).
21 PRO MAF 32/1012 (84), MAF 32/1012 (86).
22 PRO MAF 32/321 (144).
23 Leighton, PRO MAF 32/798 (39); Godney, MAF 32/135 (329); Charlton-on-Otmoor, MAF 32/910 (178); Rhosili, MAF 32/1222 (65).
24 PRO MAF 32/350 (128).
25 Charlton-on-Otmoor, PRO MAF 32/910 (178); Gotham, MAF 32/358 (153); Buckden, MAF 32/1120 (397).
26 British Portland Cement Company, PRO MAF 32/813 (23), MAF 32/751 (47); London Brick Company, MAF 32/751 (47); Stanton Iron Works, MAF 32/351 (196); Shipstones Brewery, MAF 32/373 (168); Hansons Brewery, MAF 32/356 (141); LMS Railway, MAF 32/350 (128), MAF 32/793 (11).
27 PRO MAF 32/350 (128), MAF 32/367 (161), MAF 32/373 (168).
28 PRO MAF 32/813 (23), MAF 32/1238 (80), MAF 32/1006 (22).
29 PRO MAF 32/350 (128), MAF 32/793 (11).
30 PRO MAF 32/442 (113).
31 PRO MAF 32/751 (47), MAF 32/954 (43), MAF 32/1004 (138).
32 PRO MAF 39/114 EI 739A Serial No. 659.
33 We are also grateful to Donald Sykes (personal communication, 4 November 1995), formerly Senior Lecturer in Agricultural Economics at Wye College, for further information on the use of the records at the college soon after the war.
34 PRO IR 58/29198–29200.
35 We are grateful for a personal communication on this matter from Professor J.T. Coppock, 14 September 1995.
36 Information and quotation from interview with the late Nigel Harvey, 1995.
37 The Parishes in which they were returned were South Heighton, Ditchling, Falmer, Lullington (Polegate), West Firle, Kingston (near Lewes), Berwick and Alfriston, Maresfield (two), Storrington, Upper Beeding (Shoreham), Washington, Bignor, East Dean, Coombes, Findon, Rodmell (two), Compton (Marden), Pulborough and Hartfield (three).
38 PRO MAF 32/68/137 (8), MAF 32/258/127 (70), MAF 32/55/297 (26).
39 PRO MAF 32/011/148 (550).
40 PRO MAF 32/283/86 (11), MAF 32/180/181 (1).
41 PRO MAF 32/245/165 (16).
42 PRO MAF 38/207 (14).
43 PRO MAF 38/207 (24).
44 PRO MAF 38/407.
45 For a full examination of these records and many others available in the PRO, in the context of the impact of the military on the agricultural landscape of Britain, see A.W. Foot, 'The impact of the Second World War on the agricultural landscape of Britain', unpublished M.Phil. thesis, University of Sussex, 1999.
46 PRO MAF 32/998/24 (5 and 10).
47 PRO MAF 32/1004/138 (6); 1002/48 (1); 1000/151 (3).
48 PRO MAF 32/1000/138 (46, 67); 1006/22 (4); 996/158 (1); 1015/99 (9); 1003/91 (1); 1012/86 (4); 1006/98 (11); 1013/118 (1); 442/113 (27).
49 PRO MAF 32/1000/61 (5); 1007/55 (1); 1010/148/37 (17); 1012/111 (17); 1010/103 (8); 1003/90 (15).
50 PRO MAF 32/1005/96 (16); 1006/98 (7); 1008/24 (79); 1010/148 (31); 1005/97 (21); 1015/26 (24).
51 PRO MAF 32/1001/154 (16); 1012/109 (6); 1010/102 (1); 1012/86 (7); 996/158 (11); 1033/277 (1); 1013/118 (2).

52 PRO MAF 32/994/16 (1); 995/17 (4); 367/161 (1).
53 PRO MAF 32/1008/24 (61, 800); 17/227 (7); 476/240 (13). Also see P. Wright, *The Village that died for England* (1996).
54 PRO MAF 32/1013/121 (1); 1015/124 (2, 3); 1011/45 (4).
55 See, for example PRO MAF 73/41/067; 41/16; 42/51 (East and West Sussex) or 62/40 (Pembrokeshire). See also MAF 48/394 for military training areas on the South Downs.
56 PRO MAF 73/2/14 (Berkshire); 3/46 (Bucks); 10/18 (Devon); 13/63 and 72 (Essex); 14/50 (Glocs); 119/12 and 13 and 17 (Hunts); 33/27 (Oxon); 40/37 (Surrey); 43/38 (Warwicks); 57/4 (Flints).
57 PRO MAF 32/1008/24 (95); 1015/124 (12); 1004/19 (94); 1012/109 (42).
58 PRO MAF 32/1265/86 (28, 40); 1006/22 (109); 1006/98 (13); 1002/48 (20); 1003/91 (4); 1004/38 (65); 1000/153 (31, 50); 1005/96 (24); 1008/24 (90); 1010/148 (850); 1047/43 (50).
59 PRO MAF 32/998/46 (7); 1010/148 (1); 1012/109 (33); 994/16 (9).
60 This emphasis is clear in the preface to the Summary Report written by Tom Williams, Minister of Agriculture: 'Most of the information thus collected has now served its immediate purpose – that of increasing production. But there was always in mind a second purpose for the survey. We badly needed a permanent record of all the facts and figures relating to our farming ...' in Ministry of Agriculture and Fisheries, *National Farm Survey of England and Wales (1941–1943): a Summary Report* (MAF, 1946), iii.
61 In fact the plough-up continued through to 1945, when it began to slacken (Martin, 1992).
62 See Chapter 2 for more detail on the policy issues involved in the plough-up campaign.
63 J.D. Godfrey, 'The ownership, occupation and use of land on the South Downs between the Rivers Arun and Adur in Sussex *c.* 1840–1940', unpublished D.Phil, University of Sussex 1999, Ch. 25. For the army training areas on the South Downs see PRO MAF 48: Land: Correspondence and Papers, and especially the map in 48/394. For material, including photographs of the South Downs at the end of the war, see East Sussex Record Office AMS 5666.
64 We are grateful to the late Nigel Harvey, personal communication.
65 This figure appeared in 'Statistics relating to the War Effort of the United Kingdom', a government White Paper published in November, 1944 (Ministry of Information, *What Britain has done*, 95).
66 PRO MAF 32/1008/138/1, 69 (Buxted); 88/6 (Alciston); 66/6 (West Dean); 19/52 (Burwash); 181/1 (Denny Lodge); 22/10 (E. Grinstead); 55/500 (Friston). Photographs of agricultural workers and Land Army personnel at Friston are in East Sussex Record Office, AMS 5666/1.
67 PRO MAF 32/99/10 (W. Firle); 38/28 (Uwchygarreg) where the field information is almost certainly that derived from the 1940 survey rather than the 1941–1943 survey.

Conclusions 8

A sense of transition, impermanence and imminent change came to coalesce uneasily with an older sense of Englishness by the later 1930s. That sense of old rural England changing, but always for the worse, so well encapsulated by Raymond Williams, was now so much more immediate with the threat of war. In the 6 years of war there came indeed fundamental and irreversible change to the farming community, such that it has been argued that the Second World War is of greater significance to the development of British agriculture than any comparable period since the Norman Conquest (Williams, 1973; Martin, 1992: 1).

The sense of change and an older order was caught by F.W. Bateson, Oxford don and statistical officer to the Buckinghamshire CWAEC (and active Fabian), in his poem 'Lines on the Buckinghamshire Parish Machinery Pools' in *The New Statesman and Nation* about wartime farming in the Vale of Aylesbury. Machinery pools were a cooperative venture between neighbouring farmers to provide implements and men at agreed rates but which might concentrate all available resources on any one of the farms to get the work done quickly.

> Fields I have loitered in, and villages
> Where once I canvassed Socialism, farms
> I recorded diligently, on private charts
> Plotting the public impact of blockade,
> In you I claim peculiar interest

(Bateson, 1946: xi)

© CAB *International* 2000. *The National Farm Survey, 1941–1943*
(B. Short, C. Watkins, W. Foot and P. Kinsman)

Actors, Settings and Relict Records

We have sought in this volume to bring together sets of actors within different rural settings, relating them in differing ways to national events and agendas as they unfolded, and set these side by side with literary commentators and civil servants from the ranks of the Ministries. All groupings had their own connections with the National Farm Survey (NFS) – all saw it in somewhat different ways, and pursued their interests with different methods. Thus the groups of committees that oversaw the establishment and functioning of the survey had one perspective; the groups within the localities had another – whether being the County War Agriculture Executive Committees (CWAECs), the farmers or simply interested onlookers, families, farmworkers and the generality of rural residents. Interest groups, tactical liaisons, hostilities and social cleavages run through the material, in the unusual setting of wartime England and Wales. The Survey has been related in turn to the wider political and economic perspective of policy-making in wartime Britain, and to the various manifestations of such policies which became contingent upon the Survey. The role and responsibilities of the CWAECs was in turn related to the Whitehall decisions on cropping and farm support practices, and these in turn were related to the very highest wartime decisions being made about shipping availabilities for food supplies compared with troop and munition carrying. And so the actors worked within agendas that were not of their own making.

Most importantly for our immediate purposes, we have also sought to relate the actors and their roles to the material fact of the surviving documentation. The context of the documentation within the overall wartime push for output was well caught by G.C. Hayter Hames, chairman of the Devon CWAEC in 1942 (Burrell *et al.*, 1947: 86–87):

> In Devon ... our farmers have doubled their acreage of arable crops, they have doubled their wheat acreage, they are growing five times more potatoes, they have doubled their sugar beet, they are running three times the number of tractors. ... They have ditched, tiled and drained tens of thousands of acres, they have maintained their head of cattle, increased the supply of milk and grown at home the majority of food for their livestock. They have laid water on to upland pastures and farm buildings, eradicated bracken, killed rats, bought silos and undertaken a score of other activities which they would not have dreamt of in the days before the war. More amazing still, they have managed to fill up, without exaggeration, at least ten times the number of forms that even they were accustomed to and nearly every form at least twice as diabolical as any invented by Government Departments before the War.

And as Henry Williamson (1967: 15) so accurately remarked:

> Not long after the war broke out the land of Britain was controlled by the County War Agricultural Executive Committees. The Committees governed mainly by forms.

We have demonstrated how this documentation can be placed within the wider issue of wartime food production and the stresses imposed on farming communities – and we have also explored the material itself in some depth. However, this latter task is only just beginning. The wide spatial variations in the quality of the documentation argue for a more detailed local approach which can profitably also be located within the prevailing social, political and economic circumstances of particular localities. Such an approach would then complement the national sample studied in this volume. It is believed that the records will in fact shed important light on the wartime farming complexes, which were moving ever faster into a 'productivist regime' (Marsden *et al.*, 1993).

Paradoxically many of these records demonstrate a level of control of farming that had not been witnessed since feudal regimes: the minutes of a Cultivation Sub-Committee of a CWAEC might resemble the rolls of a Court Baron in their insistence upon cutting thistles, mending fences, clearing ditches, hastening mowing or reducing sheep numbers (Fig. 8.1). State control via the CWAECs was, of course, political anathema to many farmers but the price stability and the guaranteed livelihoods now being promised were seen, even by Conservatives, to be ineluctably associated with such intervention (Bateson, 1946: 26, 152). Writing of the undoubted success of the CWAECs, A.G. Street in 1941 stated on behalf of the majority of the farming

Fig. 8.1. 'A special sense of public responsibility': a County War Agricultural Executive Committee (CWAEC) official directs operations (source: PRO INF 2/42, is Crown copyright and is reproduced with the permission of the Controller of Her Majesty's Stationery Office).

community that 'although there should be no farming from Whitehall nor by county committee, there must be control of farming' (Street, in Vesey-Fitzgerald, 1941: 134). The extent of such control was a subject, of course, for great debate but there can be little argument that although by 1939 the decision had been taken that stability and a permanent agricultural policy must be sought by state intervention, the next 6 years put in place a network of control mechanisms at local and national scales that effectively became the apparatus for a permanent post-war policy. The paper plans of Whitehall were translated into real action on some 300,000 farms, wartime practices became institutionalized and habits formed quickly (Winnifrith, 1962: 27; Smith, 1990: 88).

The documents undoubtedly helped in the wartime construction of governmental knowledge, and the place of the NFS within the wider modernist surveillance culture has already been indicated. In 1942 C.S. Orwin, Director of the Agricultural Economics Research Institute at Oxford since 1913, wrote of the need for an investigation that would include 'every aspect of rural social and industrial life' (Orwin, 1942: 58–59):

> There is no exact precedent or method established for such a survey; social surveys, soil surveys, land utilization surveys, farm management surveys, have been made, here and there, and where they are available the information they contain would be helpful.

But when Orwin, an enthusiast for land nationalization, applied through the University of Oxford to the Ministry of Agriculture for facilities to undertake such an 'experimental reconstruction survey' he was refused, despite (or because of?) the fact that he had acknowledged that much of the information already existed in the hands of the CWAECs (Orwin, 1942: 67).

Through the NFS 'local knowledges' were made available within national networks of surveillance (Matless, 1992). F.W. Bateson (1946: 164) felt that on the whole it was the officials on the CWAECs who had taken the initiative in the collection of knowledge and in its subsequent application to change farming in England:

> To some of these men the war gave opportunities that they never had before. The Executive Officers, Deputy Executive Officers, and District Officers of the active Committees have been given a reasonable degree of power and responsibility. It is a fact that the officials have seized their opportunities with two hands, whereas on the whole the Committee members have not. Without accepting the entire gospel of the 'managerial revolution', it is clear that the WAECs have proved a fruitful breeding-ground for the new type of civil servant – energetic, self-confident, full of a special sense of public responsibility.

The officials, the career civil servants and local government professionals, now ensured that agriculture became more than just a local phenomenon, rooted in its locality. It now became a 'national farm' with individual farmers united through the collection and analysis of statistics on the part of government, and rendered highly visible through the newsprint and film media (Thorpe and Pronay, 1980). Its national importance in wartime and in the difficult post-war years was acknowledged. Indeed, such was the demand for statistical information in dealing with the complexities of British farming that the ministry had probably the largest statistical organization of any department of a comparable size by the early 1960s. The NFS, seen in these terms, was one device in the creation of a 'fictive space' known as the 'national farm' (Murdoch and Ward, 1997). Entry into the Common Agricultural Policy (CAP) and the subsequent transformation of the farming sector from a productivist regime to a more extensive agriculture in the last 20 years has increased the need for monitoring and control over quotas, payments and regulations, exemplified perhaps by the Integrated Admission and Control System (IACS) introduced in 1993 throughout the European Union (Haines-Young and Watkins, 1996: 34–36).

The Survey: a Final Assessment

As a stimulus to further discussion on the purpose, role and achievement of the NFS, seen as part of the State's wartime and post-war control of agriculture, the following points can be made.

1. The completion of the NFS was a great achievement. It could, perhaps, only have been contemplated in wartime. Its purpose, however, was not clear to those who had the task of carrying out the survey work, at a time when the main attention of the CWAECs was focused on the need for greater food production.

2. The Advisory Economists had wanted a national agricultural survey for many years before the war, and took the opportunity of wartime national needs and the administrative structure that was in place to force through such a survey, whatever the costs to the wartime work. It is interesting to see the issues that were seized upon by the economists as they began to analyse the survey data, and the extent to which the modernization of farming runs as a *leitmotif* through the published material. The concerns are with electrification or mechanization, and perhaps especially with the costs of farm fragmentation, which merited a special study based on data for 24 counties, and as demonstrated in the work of Wyllie at Wye College (Whitby, 1946: 337; Wyllie, 1946). The contemporary French issues of consolidation of fields and holdings, of *remembrement*, which proceeded apace after the war in northern France, clearly had early parallels in England and Wales.

3. The really important and literally ground-breaking farm survey was that of 1940, which had a direct input into the plough-up campaign and the drive for increased food production. This was the survey the CWAECs needed to set up their programmes of work to monitor the farms and ration materials and animal feedstuffs. It is clear that CWAECs such as Buckinghamshire, Lancashire or Hampshire had their own farm survey and monitoring systems that were quite independent of those of the Ministry of Agriculture. Many of the problems of the NFS lay in the need to express these county systems by the rules (often poorly planned and ill-expressed) which aimed at a national conformity of data. It is interesting in this context to note that several publications, such as *Farmers Weekly* and even the Ministry of Information's *Land at War*, purporting to be the official story of British farming 1939–1945, refer to the 1940 survey as actually *being* the NFS (the Second Domesday), and comment on its value for wartime food production. At Uwchyrreg, Montgomeryshire, many of the primary returns give a date as July 1940, indicating clearly that the NFS, here at least, was based on the 1940 survey.

4. The fact that no proper budget for the NFS was drawn up and presented to Treasury indicates that it was ill-conceived and planned, and rapidly got out of hand, being demand-led rather than controlled. Expenditures that should have been anticipated were not even imagined until their need arose. Originally it had been thought that virtually the whole survey could be carried out by the Advisory Economists with minimal extra work thrown on the CWAECs.

5. Did the results justify all the work and expense that were put into the NFS? To answer this question, the integrity of the survey and the use to which it was eventually put have to be considered.

First, our study of the NFS data shows that they are seriously flawed. The standards by which the qualitative assessments were made vary enormously from district to district, in particular with regard to the managerial gradings. We should remember the fact that although the District Committees responsible for the gradings might contain the most appropriate people available, their selection via the CWAEC system also threw the door open to 'jobbery, favouritism [and] nepotism' (Winnifrith, 1962: 27). Wentworth Day, after his travels through wartime East Anglia, wrote of the 'mushroom bureaucracy which has grown like fungi since the war on the structure of the War Agricultural Committees [who] have their own hired trumpeters to publicise their self-conscious good deeds in the Press and in the clotted-cream accents of the BBC' (Wentworth Day, 1943: 5). But they were part-time volunteers, costing the country very little, and working on an ad hoc basis. Anthony Hurd (1951: 47), farming near Marlborough (Wiltshire), noted that:

> It was really extraordinary, even in wartime, to expect a farmer to report, however confidentially, on the personal failings of a neighbour. 'Not too energetic', 'needs an alarm clock', 'won't listen to advice because he always knows best', 'visits too many markets', 'should retire altogether', 'needs a sensible wife' – all these would be accurate comments on 'C' farmers, but, as none of us is perfect, such criticism of neighbours was rather a waste of time. It was not taken very seriously by anyone.

The general conformity of the data arising from the committees' inter-
actions with the farmers themselves (its completeness, quality and the differing interpretations of the various questions) also varies greatly since, as
Angus Calder noted 'The statistics so produced therefore represented the
opinions of various observers rather than unquestionable fact, and there
were those who were very ready to question the classification of farmers
according to merit ...' (Calder, 1969: 426). At the time that the 15% sample
of records was being selected for the national analysis, many of the farm
records (an estimated 20%) had still not been assembled. The sample for
analysis was also obtained by a matching of the Primary and Census Returns
made *after* rather than *before* selection, thereby to some extent placing the
emphasis on the good records that could be matched rather than the poor
ones that could not be so handled.

Secondly, the use of the Primary Record was severely restricted by the
confidentiality of the information it contained on the management gradings.
It is doubtful if any of the information it contained was of direct use to the
CWAECs' control of wartime farming, bearing in mind that the CWAECs had
their own systems for this which operated entirely separately from any of
the controls indicated by the NFS. The fact that there was a NFS, however,
will have stimulated many of the CWAECs to improve the quality of the data
they were keeping on the farms in their districts, as can be seen in the
examples of Buckinghamshire and East Sussex, who extended their usual
farm monitoring systems through the medium of the NFS.

One of the main reasons for the NFS was given as its future role in
post-war agricultural planning, and there is evidence that in some Advisory
Centre Provinces certain of the records were indeed used for this purpose –
in particular by the National Agricultural Advisory Service (set up in 1946)
and the Agricultural Land Service (1948).[1] It is unlikely, however, that more
than a small percentage of the overall Primary and Census Records for the
whole of England and Wales were used for such planning purposes. A
notable exception is the East Sussex Development Plan, which devised an
index of farm conditions based on accessibility, condition of buildings, convenience of layout, and provision of water and electricity for each parish in
the county (East Sussex County Council, 1953). Statistics from the national
analysis, published in the Summary Report, will also have been of considerable value. It was the farm boundary maps, however, that were most used
by the post-war planners, and this is reflected in the fact that they were
much copied and used until the 1960s – long after the individual farm
records had reverted to their archival role as the 'Second Domesday'.

Whatever criticisms may be directed against the NFS, in questioning the
need for it and the way it was carried out, we must be grateful that it did
take place, and that it was seen through to a state of completeness that
makes its survival at the Public Record Office (PRO) today such an enormous store of information for future researchers. With all its inadequacies,

errors, inconsistencies and incompleteness, it remains a most amazing body of records – a considerable achievement that reflects great credit on all who were involved in its creation. As part of our common heritage, it can now be seen as a national asset – indeed the 'Second Domesday' by which it was trumpeted from its very inception.

Research Questions

There are many ways in which this huge body of material can be exploited, and much awaits the ingenuity of future scholars to see innovative methods of approach.

One analysis of the NFS records that is certainly required is to examine the administrative units of the CWAECs and the Advisory Centres, in order to quantify more fully the regional differences in the quality of the farm survey data. To what extent are the records within one county consistent, or do they vary from district to district? Are the records consistent within the Advisory Centre Provinces, or are the differences significant from county to county? Indeed, fundamentally what were the areas covered by individual surveyors and by the Advisory Centre Recorders?

We also need to know more about the comprehensiveness of the NFS. What percentage of agricultural land in any one parish was not covered by the survey? In other words, how much was in holdings of 5 acres or less, or how much was classified as woodland? Are the maps complete? Do they show all the farms included in the individual farm records?

The spatial coverage of the survey apart, we also need to know rather more about the records themselves. In an average parish, how many Primary Returns are not matched with 4 June Returns? How many Primary Returns are without code references? How many of the 4 June 1941 Returns relate to holdings of less than 5 acres? The lower limit for inclusion within the Returns was theoretically 1 acre, having been 0.25 acres before 1892. It was estimated that about 70,000 holdings would be excluded from the survey as being between 1 and 5 acres in size, although they comprised less than 1% of the total area of crops and grass. And how many separate Census Returns are there that relate to combined holdings?

In the interpretation of the records, one immediate question that arises because of its great contemporary sensitivity was that of the accuracy of the 'A', 'B' and 'C' gradings. What evidence is there for confusion and variance in the definition of a combined holding? In general, how well were the questions on the page 3 Supplementary Form answered by the Crop Reporters?

In turning to the contemporary analysis of the survey data, it would be of interest to know whether the records that were used for the 15% sample for the national analysis can be identified, and how representative these are of the farm records within the parishes to which they relate. To what extent was it a true random sample? In the national sample undertaken for the purposes of the present research, one farm in Wiston, West Sussex, had its Primary Return marked 'Urgent. Needed for sample', presumably to allow the farm to be included in the 15% sample, and a holding in St Just in Roseland, Cornwall, included coding instructions for its similar inclusion.

Were the provincial research projects on the NFS completed and were they published? What happened to the copies of the farm boundary maps at the 2½-inch scale? And has any of Dudley Stamp's cartographical analysis of the farm survey data survived?

Finally, to what extent do the records of the CWAECs and the Advisory Centres survive? The minutes of the CWAEC (and from 1947 CAEC) meetings, their sub-committees and District Committees do survive in the PRO.[2] They will undoubtedly provide a fascinating insight into the human costs of the war to farmers, concerned as the minutes are with cultivation orders and dispossessions, as well as the plethora of other tasks allocated to the CWAECs to undertake. Indeed, the minutes will provide the first full picture of the rural social conflicts arising in wartime, highlighted by the tragic George Walden case at Itchen Stoke, Hampshire, but echoed many times elsewhere (Mountford, 1997).

We might also add that the role of the main national committee members is also of interest (see Appendix). Dudley Stamp, such an powerful influence in the development of Geography as a discipline, moves between the National Land Use Survey during the 1930s and the NFS Research Sub-Committee in the 1940s. In his preface to the third edition of *The Land of Britain: its Use and Misuse* (Stamp, 1962: 1) he refers to the fact that this summary volume was able to include 'an account of wartime changes and the approach to post-war national planning through events in which I myself played some part'. This modest phrase can now be amplified to some extent at least by reference to the minutes of the committees on which he worked during the planning of the NFS, a role he does not himself mention even though he does devote a paragraph to the survey, and reproduces copies of NFS forms (Stamp, 1950: 407–411).[3]

Students of rural Britain have long envied the cadastral surveys of French, Swedish and other colleagues. But for 1941–1943 at least, we now have a comparable source which similarly serves the interests of the State (Kain and Baigent, 1992). As well as gaining access to the 4 June 1941 Returns, we are able to clothe the statistics with qualitative information on the farm and its occupier, on the socio-political context of the survey as seen by senior civil servants, and with the spatial information drawn from the large-scale maps used in the survey.

In its detail and coverage the NFS has no precise equal in Britain. It is less comprehensive in coverage spatially than the Lloyd George 'Domesday' 1910, since the latter covered all hereditaments, urban and rural and irrespective of size, but the quality of information at the farm level is incomparably greater for 1941–1943. It is also of more widespread relevance, and has more information than the tithe surveys of the 1840s which, while good in coverage for the South East and Wales, are less good for the Midlands or the North (Kain and Prince, 1985; Short, 1997). However, where the information is extant for all three dates we now have benchmarks for the 1840s, 1910 and 1940s against which to measure many aspects of farming and rural change, not least of which will be detailed studies of land ownership and farming structure. Such studies will now, 50 years on, bring about the research originally envisaged by Tom Williams, Minister of Agriculture and Fisheries, who wrote in the preface to the 1946 Summary Report of the NFS (Ministry of Agriculture and Fisheries, 1946: iii):

> No such record need merely lie in the archives of a Department: it can also be made available to the general public … and to scholars in our Universities and Research Institutions for further study and elaboration.

Notes

1   Personal communication from Donald Sykes (September and November 1995), formerly Senior Lecturer in Agricultural Economics, Wye College. Sykes remembers that among others consulting the NFS material in the early 1950s was R. Green (Rothamsted), author of the Soil Survey monograph on Romney Marsh, and J. Wyllie, Provincial Agricultural Economist at Wye College until 1951.
2   PRO MAF 80. The records were originally closed for a 50-year period but the closure was lifted in 1997. The records contain material relating to the whole period 1939–1957, arranged in alphabetical order of English and Welsh counties. Many of the volumes are indexed (e.g. MAF 80/810, Gloucestershire Executive Committee minutes 19 May 1939, to 1 March 1940).
3   Stamp reproduced the 4 June Return form using the Form 398/SS rather than the C47/SSY which is more normally preserved in the PRO MAF 32 class. In some MAF 32 records, both versions are to be found, e.g. Tyneham, Dorset (PRO MAF 32/476 (240)).

Appendix

A list of the Provincial Agricultural Advisory Centres drawn up in 1948 (PRO MAF 38/867) shows that there were at that date ten provinces, whereas at the time of the National Farm Survey there had been 11. There had, in fact, been several changes in the allocation of counties to the provinces, as well as various address changes of the Advisory Centres. The Harper Adams College, Newport (the Advisory Province for Shropshire, Stafford and Warwickshire) had disappeared from the list, and its counties given to the Southern province (Reading) and the North Western province (Manchester).

The Advisory Economists

Northern Province
    Cumberland
    Westmorland
    Northumberland

D.H. Dinsdale,
King's College,
Newcastle-on-Tyne

Yorkshire
    Yorkshire, E. Riding
    Yorkshire, N. Riding
    Yorkshire, W. Riding

W.H. Long,
Department of Agriculture,
The University,
Leeds

East Midland
    Derby
    Leicester
    Lincoln (Kesteven)

M.A. Knox,
Midland Agricultural College,
Sutton Bonington,
Loughborough

Lincoln (Lindsey)
Nottingham
Rutland

Eastern   A.W. Menzies-Kitchin,
  Bedford   School of Agriculture,
  Cambridge   University of Cambridge,
  Essex   Cambridge
  Hertford
  Huntingdon
  Isle of Ely
  Lincoln (Holland)
  Norfolk
  Soke of Peterborough
  Suffolk, East
  Suffolk, West

South Eastern   J. Wyllie,
  Kent   South Eastern Agricultural College,
  Surrey   Wye,
  Sussex, East   Kent
  Sussex, West

Southern   E. Thomas,
  Berkshire   Department of Agricultural Economics,
  Buckingham   University of Reading,
  Dorset   Reading
  Hampshire
  Isle of Wight
  Middlesex
  Northampton
  Oxford

South Western   R. Henderson,
  Cornwall   Seale Hayne Agricultural College,
  Devon   Newton Abbot,
  Isles of Scilly   Devon

Western   C.V. Dawe,
  Gloucester   Agricultural Advisory Department,
  Hereford   Bristol; and University Research Station,
  Somerset   Long Ashton,
  Wiltshire   Bristol
  Worcester

West Midland   J.J. MacGregor,
  Shropshire   Harper Adams Agricultural College,
  Stafford   Newport,
  Warwick   Shropshire

North Western	A. Jones,
Cheshire	Agricultural Advisory Department,
Lancashire	The University, Manchester
Wales	A.W. Ashby,
	Agricultural Research Buildings,
	Aberystwyth; and
	A. Bridges,
	Agricultural Economics Research
	Institute, Oxford

Members of the Farm Survey Committee

This committee was set up to review the progress of the 1940 farm survey and to make recommendations on how that survey might be improved and extended: its place was taken later by the Farm Survey Supervisory Committee, set up from May 1941.

Sir Donald Fergusson (Chairman)	Ministry of Agriculture and Fisheries
Mr R.R. Enfield	Ministry of Agriculture and Fisheries
Mr Anthony Hurd	Ministry of Agriculture and Fisheries
Mr C. Bryner Jones	Ministry of Agriculture and Fisheries
Mr E.L. Mitchell	Ministry of Agriculture and Fisheries
Sir Edwin Butler	Agricultural Research Council
Major T.K. Jeans	Chairman of the Salisbury District Committee of the Wiltshire WAEC
Mr W.J. Pulford	Executive Officer of Shropshire WAEC
Mr F. Rayns	Executive Officer of the Norfolk WAEC
Dr L. Dudley Stamp	Land Utilisation Survey
Mr F. Russell Wood	Chairman of the Hitchin District Committee of the Hertfordshire WAEC
Mr M.G. Kendall (Secretary)	Ministry of Agriculture and Fisheries

Members of the Farm Survey Supervisory Committee

(First meeting on 27 May 1941)

Mr R.R. Enfield (Chairman)	Ministry of Agriculture and Fisheries
Mr S. Fitch	Ministry of Agriculture and Fisheries
Mr J.H. Kirk	Ministry of Agriculture and Fisheries
Mr A. Bridges	Agricultural Economics Research Institute
Mr D.H. Dinsdale	University of Durham, Agricultural Department
Major T.K. Jeans	Chairman of the Salisbury District Committee of the Wiltshire WAEC

Mr E. Wynne Jones	Forestry Commission
Mr F. Rayns	Norfolk WAEC
Prof. G.W. Robinson	University College of North Wales, Agricultural Department (Agricultural Research Council)
Dr L. Dudley Stamp	Land Utilisation Survey
Mr R.W. Trumper	
Prof. J.A. Scott Watson	School of Rural Economy, University of Oxford
Mr F. Yates	Rothamsted Experimental Station, Harpenden
Mr H. Whitby (Secretary)	Ministry of Agriculture and Fisheries

Members of the Farm Survey Research Sub-Committee

This sub-committee of the Farm Survey Supervisory Committee was established in March 1942 in order to evaluate the farm records being obtained, to suggest how the information gained might be best used and to develop a system for national and provincial analysis.

Mr J.H. Kirk (Chairman)
Major T.K. Jeans
Dr L. Dudley Stamp
Mr Edgar Thomas
Dr F. Yates
Mr H. Whitby (Secretary)

Members of a Sub-Committee of the Conference of Advisory Economists

First meeting relating to the National Farm Survey held on 16 October 1940 – a further Sub-Committee met on 6 February 1941.

Mr R.R. Enfield (Chairman)
Mr M.G. Kendall
Prof. A.W. Ashby
Mr A. Bridges
Dr C.V. Dawe
Mr Arthur Jones
Dr A.W. Menzies-Kitchin
Mr Edgar Thomas
Mr H. Whitby

Individuals Present at a Special Meeting of the Conference of Advisory Economists – 6 November 1940

Mr R.R. Enfield (Chairman)
Mr M.G. Kendall
Prof. A.W. Ashby
Mr A. Bridges
Dr C.V. Dawe
Mr D.H. Dinsdale
Mr R. Henderson
Mr Arthur Jones
Mr M.A. Knox
Dr A.W. Menzies-Kitchin
Mr W.H. Long
Mr J.J. MacGregor
Mr Edgar Thomas
Mr J. Wyllie
Mr H. Whitby (Secretary)

References

Anon. (nd) *A Norfolk Woman: Farming on a Battleground*. Geo. R. Reeve, London.

Astor, Viscount and Murray, K.A.H. (1933) *The Planning of Agriculture*. Oxford University Press, Oxford.

Astor, Viscount and Rowntree, B.S. (1938) *British Agriculture: the Principles of a Future Policy*. Longmans, London.

Baldwin, C.M. (nd) *Digging for victory*. Unpublished manuscript, Rural Life Centre, University of Reading.

Barrell, J. (1972) *The Idea of Landscape and the Sense of Place, 1730–1840. An Approach to the Poetry of John Clare*. Cambridge University Press, Cambridge.

Bateson, F.W. (1940) *Mixed Farming and Muddled Thinking*. MacDonald, London.

Bateson, F.W. (1946) The Democratization of Control. In: Bateson, F.W. (ed.) *Towards a Socialist Agriculture: Studies by a Group of Fabians*. Victor Gollancz, London, pp. 151–168.

Bateson, F.W. (ed.) (1946) *Towards a Socialist Agriculture: Studies by a Group of Fabians*. Victor Gollancz, London.

Blishen, E. (1972) *A Cack-handed War*. Thames and Hudson, London.

Bowler, I. (1986) Government Agricultural policies. In: Pacione, M. (ed.) *Progress in Agricultural Geography*. Croom Helm, London, pp. 124–148.

Brown, J. (1989) *Agriculture in England: a Survey of Farming, 1870–1947*. Manchester University Press, Manchester.

Burrell, M.R., Cornwallis, C. and Hayter Haymes, G.C. (1947) War-time food production: the work of the Agricultural Executive Committees. *Journal of the Royal Agricultural Society of England* 108, 70–90.

Calder, A. (1969) *The People's War: Britain 1939–45.* Jonathan Cape, London.

Cantwell, J.D. (1993) *The Second World War: a Guide to Documents in the Public Record Office.* HMSO, London.

Cooke, G.W. (ed.) (1981) *Agricultural Research 1931–1981.* Agricultural Research Council, London.

Cooper, A.F. (1989) *British Agricultural Policy, 1912–36: a Study in Conservative Politics.* Manchester University Press, Manchester.

Coppock, J.T. (1955) The relationship of farm and parish boundaries – a study in the use of agricultural statistics. *Geographical Studies* 2, 12–26.

Coppock, J.T. (1960a) Crop and livestock changes in the Chilterns, 1931–51. *Transactions of the Institute of British Geographers* 28,179–198.

Coppock, J.T. (1960b) Farms and fields in the Chilterns. *Erdkunde* 14, 134–146.

Coppock, J.T. (1960c) The parish as a geographical–statistical unit. *Tijdschrift voor Economische en Sociale Geographie* 51(12), 317–326.

Cox, G., Lowe, P. and Winter, M. (1985) Changing directions in agricultural policy: corporatist arrangements in production and conservation policies. *Sociologia Ruralis* 25, 130–154.

Cox, G., Lowe, P. and Winter, M. (1986) *Agriculture: People and Policies.* Allen and Unwin, London.

Cox, G., Lowe, P. and Winter, M. (1991) The origins and early development of the National Farmers' Union. *Agricultural History Review* 39(1), 30–47.

Cox, P.W. (1944) Front-Line Farming: Kent's war-time effort. *Agriculture* 51(3), 118–123.

Dale, H.E. (1939) Agriculture and the Civil Service. In: *Agriculture in the Twentieth Century: Essays on Research, Practice and Organisation to be presented to Sir Daniel Hall.* Oxford University Press, Oxford, pp. 1–20.

Dewey, P.E. (1989) *British Agriculture in the First World War.* Routledge, London.

Douet, A. (1990) Norfolk agriculture 1918–1972. Unpublished PhD Thesis, University of East Anglia.

Douet, A. (1996) Some aspects of Sugar Beet production in England, 1945–1985. *Rural History* 7(2), 221–238.

Easterbrook, L.F. (1943) *British Agriculture: British Life and Thought, No.16.* British Council/Longmans Green and Co., London.

East Sussex County Council (1953) *East Sussex County Council Development Plan.*

Enfield, R. (1924) *The Agricultural Crisis 1920–23.* Longmans, London.

Farmers' Rights Association (1945a) *Living Casualties (the Dispossessed Farmer): an Account of Bureacratic Control of Agriculture Exercised through War Agricultural Executive Committees.* FRA, Church Stretton, Salop.

Farmers' Rights Association (1945b) *The New Morality*. FRA, Church Stretton, Salop.

Farmers' Rights Association (1948) *The New Anarchy*. FRA, Church Stretton, Salop.

Finney, D.J. and Yates, F. (1981) Statistics and Computing in Agricultural Research. In: Cooke, G.W. (ed.) *Agricultural Research 1931–1981*. Agricultural Research Council, London, pp. 219–236.

Fitzrandolph, H.E. and Hay, M.D. (1926a) *The Rural Industries of England and Wales*, Vol. I, *Timber and Underwood Industries and Some Village Workshops*. Oxford University Press, Oxford.

Fitzrandolph, H.E. and Hay, M.D. (1926b) *The Rural Industries of England and Wales*, Vol. II, *Osier-growing and Basketry and Some Rural Factories*. Oxford University Press, Oxford.

Fitzrandolph, H.E. and Hay, M.D. (1927a) *The Rural Industries of England and Wales*, Vol. III, *Decorative Crafts and Rural Potteries*. Oxford University Press, Oxford.

Fitzrandolph, H.E. and Hay, M.D. (1927b) *The Rural Industries of England and Wales*, Vol. IV, *Rural Industries in Wales*. Oxford University Press, Oxford.

Foot, A.W. (1994) *Maps for Family History: a Guide to Records of the Tithe, Valuation Office, and National Farm Surveys of England and Wales, 1836–1943*. Public Record Office Readers' Guide No.9, PRO, London.

Foot, A.W. (1999) The impact of the military on the British farming landscape in the Second World War. Unpublished MPhil Thesis, University of Sussex.

Foreman, S. (1989) *Loaves and Fishes: an Illustrated History of MAFF 1889–1989*. HMSO, London.

Fryer, D.W. (1942) *Land of Britain: Buckinghamshire, Part 54*. Longman, Green and Co., London.

Gardiner, R. (1943) *England Herself: Ventures in Rural Restoration*. Faber and Faber, London.

Garrad, G.H. (1954) *A Survey of the Agriculture of Kent*. Royal Agricultural Society of England. Rase, London.

Gasson, R. (1980) *The Role of Women in British Agriculture*. Women's Farm and Garden Association, Colchester.

Godfrey, J.D. (1999) The ownership, occupation and use of land on the South Downs between the Rivers Arun and Adur in Sussex *c.*1840–1940. Unpublished DPhil Thesis, University of Sussex.

Haines-Young, R. and Watkins, C. (1996) The rural data infrastructure. *International Journal Geographical Information Systems* 10(1), 21–46.

Hall, A.D. (1941) *Reconstruction and the Land: an Approach to Farming in the National Interest*. Macmillan and Co., London.

Hammond, R.J. (1951) *Food*, Vol. I, *The Growth of Policy*. Longman, Green and Co./HMSO, London.

Hammond, R.J. (1954) *Food and Agriculture in Britain 1939–45: Aspects of Wartime Control*. Stamford University Press, Stamford, California.

Hammond, R.J. (1956) *Food*, Vol. II, *Studies in Administration and Control*. Longman, Green and Co./HMSO, London.

Hammond, R.J. (1962) *Food*, Vol. III, *Studies in Administration and Control*. Longman, Green and Co./HMSO, London.

Havinden, M.A. (1966) *Estate Villages*. Lund Humphries, Reading.

Henderson, G. (1944) *The Farming Ladder*. Faber and Faber, London.

Henderson, G. (1950) *Farmer's Progress*. Faber and Faber, London.

Hinckes, R.T. (1913) *The Farmer's Outlook. A Review of Home and Overseas Agriculture 1880–1913*. Jarrold, London.

Hurd, A. (1951) *A Farmer in Whitehall: Britain's Farming Revolution 1939–1950 and Future Prospects*. Country Life, London.

Kain, R.J.P. and Baigent, E. (1992) *The Cadastral Map in the Service of the State*. Chicago University Press, Chicago.

Kain, R.J.P. and Prince, H.C. (1985) *The Tithe Surveys of England and Wales*. Cambridge University Press, Cambridge.

Kendall, M.G. (1939) The geographical distribution of crop productivity in England. *Journal Royal Statistical Society* 102, 21–48.

Kirk, J.H. (1979) *The Development of Agriculture in Germany and the UK. No. 2, UK Agricultural policy 1870–1970*. Centre for European Agricultural Studies, Wye College, Kent.

Long, W.H. and Davies, G.M. (1948) *Farm Life in a Yorkshire Dale: an Economic Study of Swaledale*. Dalesman, Clapham, Yorkshire.

MacKenzie, D.A. (1981) *Statistics in Britain 1865–1930: the Social Construction of Scientific Knowledge*. Edinburgh University Press, Edinburgh.

Marks, H.F. and Britton, D.K. (eds) (1989) *A Hundred Years of British Food and Farming: a Statistical Survey*. Taylor & Francis, London.

Marsden, T., Murdoch, J., Lowe, P., Munton, R. and Flynn, A. (1993) *Constructing the Countryside*. UCL Press, London.

Marshall, M. (1943) *Land of Britain: Oxfordshire, Part 56*. Geographical Publications, London.

Martin, J.F. (1992) The impact of government intervention on agricultural productivity in England and Wales 1939–45. Unpublished PhD Thesis, University of Reading.

Martin, J.F. (1999) *The Development of Modern Agriculture: British Farming Since 1931*. Manchester University Press, Manchester.

Matless, D. (1992) Regional surveys and local knowledges: the geographical imagination in Britain, 1918–39. *Transactions of the Institute of British Geographers* 17, 448–463.

Matless, D. (1998) *Landscape and Englishness*. Reaktion, London.

Menzies-Kitchin, A.W. and Chapman, W.D. (1946) War-time changes in the organisation of two groups of Eastern counties farms. *Economic Journal* 56, 37–85.

Milner Committee (1915) *Interim Report of the Departmental Committee Appointed By the President of the Board of Agriculture to Consider the Production of Food in England and Wales.* Board of Agriculture, Cd. 8048, Parliamentary Papers 1914–1916, V. Final Report, Cd. 8095.

Ministry of Agriculture and Fisheries (1946) *National Farm Survey of England and Wales: a Summary Report.* HMSO, London.

Ministry of Information (1944) *The Farms Fight Too.* HMSO, London.

Ministry of Information (1945a) *What Britain has Done 1939–1945.* HMSO, London.

Ministry of Information (1945b) *Land at War: the Official Story of British Farming 1939–1944.* HMSO, London.

Moss, L. (1991) *The Government Social Survey: a History.* HMSO, London.

Mountford, F. (1997) *Heartbreak Farm: a Farmer and his Farm in Wartime.* Sutton, London.

Murdoch, J. and Ward, N. (1997) Governmentality and territoriality: the statistical manufacture of Britain's 'national farm'. *Political Geography* 16(4), 307–324.

Murray, K.A.H. (1955) *Agriculture.* History of the Second World War, United Kingdom Civil Series, Longman, Green and Co./HMSO, London.

Orr, J. (1916) *Agriculture in Oxfordshire. A Survey.* Clarendon Press, Oxford.

Orwin, C.S. (1930) *The Future of Farming.* Clarendon Press, Oxford.

Orwin, C.S. (1938) *Agricultural Economics 1913–1938.* Agricultural Economics Research Institute, Oxford.

Orwin, C.S. (1942) *Speed the Plough.* Penguin, Harmondsworth.

Parsons, L.M. (1969) Land Commissioners of England 1894. *Agriculture* 76, 454–459.

Penning-Rowsell, E.C. (1997) Who 'betrayed' whom? Power and politics in the 1920/21 agricultural crisis. *Agricultural History Review* 45(2), 176–194.

Robinson, J.M. (1989) *The Country House at War.* Bodley Head, London.

Rogers, A. (1999) *The Most Revolutionary Measure: a History of the Rural Development Commission 1909–1999.* RDC, Salisbury.

Russell, E.J. (1966) *A History of Agricultural Science in Great Britain 1620–1954.* Allen and Unwin, London.

Rutherford, R.S.G. and Bateson, F.W. (1946) Co-operation in Agriculture. In: Bateson, F.W. (ed.) *Towards a Socialist Agriculture.* Gollancz, 124–136.

Rycroft, S. and Cosgrove, D. (1995) Mapping the modern nation: Dudley Stamp and the Land Utilisation Survey. *History Workshop Journal* 40, 91–105.

Self, P. and Storing, H.J. (1962) *The State and the Farmer.* George Allen and Unwin, London.

Shaw, M. and Miles, I. (1979) The social roots of statistical knowledge. In: Irvine, J., Miles, I. and Evans, J. (eds) *Demystifying Social Statistics.* Pluto, London, pp. 27–38.

Shawyer, A.J. (1990) Farm structure and farm families: a Nottinghamshire field area. *East Midland Geographer* 13(1), 1–18.

Short, B. (1997) *Land and Society in Edwardian Britain.* Cambridge University Press, Cambridge.

Short, B. and Watkins, C. (1994) The National Farm Survey of England and Wales, 1941–3. *Area* 26(3), 288–293.

Smith, M.J. (1990) *The Politics of Agricultural Support in Britain: the Development of the Agricultural Policy Community.* Dartmouth, Aldershot.

Stamp, L. Dudley (1947) Wartime changes in British Agriculture. *Geographical Journal* 109, 39–57.

Stamp, L. Dudley (1947) (2nd edn 1950, 3rd edn 1962) *The Land of Britain: its Use and Misuse.* Longman, Green and Co./Geographical Publications, London.

Stapledon, R.G. (1935) *The Land Now and To-morrow.* Faber and Faber, London.

Stapledon, R.G. (1939) *The Plough-up Policy and Ley Farming.* Faber and Faber, London.

Stapledon, R.G. (1943) *The Way of the Land.* Faber and Faber, London.

Street, A.G. (1941) Untitled contribution. In: Vesey-Fitzgerald, B. (ed.) *Programme for Agriculture.* Michael Joseph, London, pp. 131–139.

Street, A.G. (1943) *Hitler's Whistle.* Eyre and Spottiswoode, London.

Street, A.G. (1954) *Feather-bedding.* Faber and Faber, London.

Sykes, F. (1944) *This Farming Business.* Faber and Faber, London.

Thomas, E. and Elms, C.E. (1938) *An Economic Survey of Buckinghamshire Agriculture, Part I.* Faculty of Agriculture and Horticulture, University of Reading.

Thorpe, F. and Pronay, N. (1980) *British Official Films in the Second World War.* Clio Press, London.

Venn, J.A. (1933) *The Foundation of Agricultural Economics: together with an economic history of British agriculture during and after the Great War.* Cambridge University Press, Cambridge.

Venn, J.A. (1939) Agriculture and the State: the financial and economic results of control. In: *Agriculture in the Twentieth Century: Essays on Research, Practice and Organisation to be Presented to Sir Daniel Hall.* Oxford University Press, Oxford, pp. 21–49.

Vesey-FitzGerald, B. (ed.) (1941) *Programme for Agriculture.* Michael Joseph, London.

Waller, R. (1962) *Prophet of the New Age: the Life and Thought of Sir George Stapledon, FRS.* Faber and Faber, London.

Watkins, C. (1984) The use of Forestry Commission censuses for the study of woodland change. *Journal of Historical Geography* 10, 396–406.

Watkins, C. (1985) Sources for the assessment of British woodland change. *Applied Geography* 5, 153–168.

Waugh, E. (1945) *Brideshead Revisited: the Sacred and Profane Memories of Captain Charles Ryder: a Novel.* Chapman and Hall, London.

Webber, A. (1982) Government policy and British agriculture 1917–1939. Unpublished PhD Thesis, University of Kent.

Wentworth-Day, J. (1943) *Farming Adventure: a Thousand Miles through England on a Horse.* Harrap, London.

Wentworth-Day, J. (1950) *Marshland Adventure.* Harrap, London.

Whatmore, S. (1991) *Farming Women: Gender, Work and Family Enterprise.* Macmillan, London.

Whetham, E.H. (1952) *British Farming 1939–49.* Thomas Nelson and Sons, London.

Whetham, E.H. (1974) The Agriculture Act 1920 and its repeal – the 'Great Betrayal'. *Agricultural History Review* 22, 36–49.

Whetham, E.H. (1978) *The Agrarian History of England and Wales,* Vol. VIII, *1914–39.* Cambridge University Press, Cambridge.

Whitby, H. (1946) National Farm Survey of England and Wales. *Agriculture* 53(8), 335–340.

Williams, R. (1973) *The Country and the City.* Chatto and Windus, London.

Williamson, H. (1967) *Lucifer before Sunrise.* MacDonald, London.

Winnifreth, Sir John (1962) *The Ministry of Agriculture, Fisheries and Food* (New Whitehall Series No. 11). HMSO, London.

Winter, M. (1996) *Rural Politics: Policies for Agriculture, Forestry and the Environment.* Routledge, London.

Wood, B. (1982) The development of the Nottinghamshire County Council smallholdings estate, 1907–1980. *East Midland Geographer* 8, 25–35.

Wright, P. (1996) *The Village that Died for England: the Strange Story of Tyneham.* Cape, London.

Wyllie, R.T. (1946) *National Farm Survey: Wye Province.* Wye College, Kent.

Index

Figures in **bold** indicate major references.
Figures in *italic* refer to diagrams, photographs and tables.

A Cackhanded War (Blishen) 47
acreage 66, 114, 116, **171–175**, *173*, *174*
    maps 148–149, *149*
    ploughed-up land 218, 219
Advisory Centres 47–50, *48*, *51*, 52,
        69–70, 72, 95
    Aberystwyth 62, 89
    Bristol 89, 106
    Cambridge 89
    Harper-Adams 89
    Leeds 90
    Manchester 62, 90
    Midlands 90
    Newcastle 90
    Reading 90, 110
    Seale Hayne 49, 62, 90
    Wye 49, 73, 91
Advisory Economists 22, 24, **47**, 49, *51*,
        58, 233
    access to farm codes 71–72
    coding instructions 87
    collation of data 55
      maps 67, 68, 76, 107–108
      Primary Return 52, 64–65
    data analysis 5, 81, 93, 140
    details of survey 60, 62, 66

    list of 56, **230–240**, 239–241
    records 110
    regional analysis 87–91
    special meeting 243
    sub-committee members 242
Agricultural Advisory Provinces 11, *48*,
        99, *115*, 235
    farm fragmentation 177
    holding size *176*
    ploughed-up land 216–218, *217*, *218*
agricultural costing 22–23, 24
Agricultural Costing Committee 22
agricultural economics 22–25, 39
Agricultural Economics Research Institute
      (AERI) 22, 23, 105, 232
Agricultural Economics Society 23
Agricultural Education Committees 38
Agricultural Holdings Acts 27, 59
Agricultural Land Service 235
Agricultural Policy Sub-Committee
      Reconstruction Committee 27
agricultural prices index 16
agricultural production policies 27
agricultural provinces 23
Agricultural Research Council 2, 25
Agricultural Returns Order 64

agriculture
  government intervention 1–2, 16, 17,
    19–20, 24, **39–40**
  centralization 27
  expenditure 20
  land management 27
  policies
    interwar 15–19
    war 37–39
  statistical surveys 1–3
    funding 23
    ideological opposition 39
    interwar 21–25
Agriculture Act (1920) 15–16
Agrostis spp. 32
America 3
animal feed 34
Annals of Agriculture 21
arable land 127
archival policies
  archive transfer 106–109
  confidentiality 104–106
  destruction 110
  preservation 58, 110, 157
  see also documents
Argentine, The 16
Australia 16

bacon 16
bad farming certificate 59
Baker, A.C. 91, 94
Baldwin, Stanley 16, 18
barbed wire *199*, 201
barley 33, 222
Bartlett, A.B. 95
basal diet 34
Bateson, F.W. 99–100, 229, 232
Battle of Britain 3
beans 33, 34
Bedfordshire 156, 163, 167, 170
Berkshire 167
Biggs, H. 95
birds 124
Black, W.R. 94
Board of Agriculture 2, 19, 25, 26, 52
Boscawen, Sir Arthur Griffith 16
boundaries 5, 54, 77, 103, 146, *178*,
  180
Brideshead Revisited (Waugh) 202
Buckinghamshire 5, 35, 99–100, 110,
  163, 219, 229, 233
built-up areas 11

bureaucracy 37, 38

C47/SSY (blue form, census) 63, 73
C49/SSY (census form) 63, 95
C51/SSY 64
C69/SSY (white form, census) 64, 97
'C' farms 42, 54
cadastral surveys 237
Calder, Angus 235
Cambridgeshire 157, 170
Canada 16
card index 49
cards, punched 82, 86
cattle 16
Census Return 52, 56, 132–135, 138
  accuracy 50, 94–95
  acreage figures *174*
  completion of 73–74
  crops and grass 133–134
  forms 63–64
  labour 134–135
  livestock 135
  military activities 200
  problems 94–97, 100
  supplementary enquiries 5
Central Statistical Office 3
cereals 15, 16
Cheshire 42, 150
Church of England 169
'circle charts' 91
Circular 545 (MAF) 3, 56
classification system
  farm management *see* managerial
    classifications
  land 59
clergy, local 169–170
clubs and societies 170
coding sheets 86
coding system
  farms 71–73, 75, 116–117
  punched cards 86
colour (maps) 75, 146–147
combine harvesters *182*
Common Agricultural Policy (CAP)
  233
confidentiality 57, 61, 96, 98, **104–106**,
  235
Cooperative Societies 170
corn 16, 27, 33
Cornwall 184, 185, 201, 237
costs 22, 24, 94, 95, 103–104, 234
cottages 124, 203

counties
    coding sheets 86
    coding systems 71, 73
    differences of grading 59
    map *44*
    results 85
    survey progress 58, 59, *61*
Central Landowners Association 20, 105
County Agricultural Executive Committees
        32
county farm record systems 38–39
County War Agricultural Executive
        Committees (CWAECs) 5, 15,
        44, **45–47**, *46, 51*, 230–232,
        231
    clerical staff 29
    closure 110
    communications with government
        30, 37, 56, 66, 102
    coordination 20
    data analysis 81
    day-to-day work 26–27, 30, 38
    establishment 25
    Executive Officers 56
    first Farm Survey 41, 42, 43
    instructions to 26
    machinery pools 181–182, *181*,
        229
    organization 25, 26, 33
        district committees 45
        key figures 30–31
        sub-committees 45
            local 29
            specialist 29–30
    plough-up campaign 206, 214, 218,
        223–224
    plough-up orders 37
    plough-up quotas 35
    post-war 31–32, 40
    powers 25–26, 28, 30
        evictions 37
        supply of information 57, 69
    records 30
    Shropshire 70
    special officers 29
    survey work 52, 55, 61, 68–69
        lack of enthusiasm 58, 99
        maps 74–75
credit scheme 17
crop acreages *36*
Crop Reporters 5, *51*
    area codes 71, 116

    payments to 103
    role of 50, 73, 95, 96–97, 103
cropping orders 38
crops sown 33, 131, 210–214, *211, 212,
        213*, 222–223
Crown Estate 169
Cultivation of Land Orders 25
Cumberland 59
Czechoslovakia 33

Dale, H.E. 19, 20
Dartmoor 179
data analysis
    cartographic 91–92
    national 81–87
    regional 87–91
data assessment 113–114
    Census Return 132–135
    Horticultural Return 135–137
    internal consistency 139–142,
        162–163, 172, *173*
    Primary Return 114–132
    Supplementary Form 137–139
data collection 11
data transcription 11, 13
databases *114*
    aim of 11–12
    national sample 11, *12, 115*
    regional samples 10–11, *12*
    table structure 11
Davies, G.M. 88
Day, Wentworth 234
De La Warr, Earl 17
Defence Regulations 30, 57, 61, 64, 69,
        104, 105
deficiency payments 15, 16, 18
Depression, the 16
Derbyshire 10, 35
    Beeley Estate 167
    Chatsworth Estates 153, 167, 168
    Forestry Commission 168, 169
    maps 146, 147, 148, 150, 151, 152,
        154
        fragmented holdings 155
        grazing rights 156
        presentation 157
        reservoirs 168
Development Fund 103
Devon 42, 102–103, 184, 230
dispossessions *see* evictions
district committees 26, 69, 71
ditches 123

documents
    access to 105–106
    Census Return *see* Census Return
    circulation *53*
    Horticultural Return *see* Horticultural
        Return
    index 49
    maps *see* maps
    Primary Return *see* Primary Return
    reminder notices *65*
    reminder form 139
    Supplementary Form *see*
        Supplementary Form
    see also archival policies
Dolfor Scheme 182, 183, 216
Domesday Book 54, 60, 106, 206
Dorman-Smith, Sir Reginald 33, 37, *207*
Durham 59, 108

East Anglia 16, 198, 221, 234
Economic and Social Research Council 10
educational institutions 170
electric motors 185, 186
electricity 91, 92, 94, 125
Elliot, Walter 16, 18, 28
Enfield, R.R. 88, 93, 95
engines 184–186
England, ploughed-up land 216, *217*, 218
Ernle, Lord 27
errors 94
Essex 26, 35, 180, 221
        maps 145, 148, 154
estates, landed 82, 165, 166–167
Eugenics movement 2
European Union 233
evictions 37, 59, 224
executors 167

family workers 134, 137
farm accounting 24
farm buildings 123, 201
farm code reference 50, 71–73
farm incomes 16–17
farm management surveys 23–24, 47
Farm Records 59
Farm Survey Committee 3, 54, 66
        members 241
Farm Survey Progress Report 61
Farm Survey Record Form 66
Farm Survey Research Sub-Committee 52,
        242
Farm Survey Supervisor 48

Farm Survey Supervisory Committee 52
    data analysis 81
    farm boundaries 68, 103
    managerial classifications 58, 59
    maps 76
    members 241–242
    publication 93
    random sampling proposal 55–56
    survey process 69
    survey progress 57–58, 61
farmers *51*, 63, 92, 119
    armed forces 205
    failings 56–57, 98
    grazing rights 122
    managerial classifications *see* man-
        agerial classifications
    other occupations 121–122
    refusal to answer questions 95, 97
Farmers' Rights Association 37
Farmers and Smallholders Association
        37
Farmers Weekly 42, 234
farmhouse 123, 125
farms
    acreage *see* acreage
    boundaries 5, 54, 77, 103, 146, *178*,
        180
    condition of 123–125
    county records 38–39
    enemy action 206
    fragmentation 60–61, *149*, 154–155,
        171, 176–180, *178*, *179*, 233
    grading *see* managerial classifications
    holdings *see* holdings
    links to Ministry *53*
    mechanization *see* mechanization
    size 171
        acreage *see* acreage
        holdings *see* holdings
    tenure *see* tenure
fences 124, 201
fertilizers 34, 39, 49, 127–128
Field Workers *see* Recorders
fields
    derelict 125
    drainage 124
    water supply 125–126
filing cabinets 50
filing clerks 49–50
filing system 49
financial institutions 170–171
first Farm Survey 1940 41–43

First World War 15, 19, 22, 24
    state involvement in agriculture 27, 38
fisheries 19
food
    calorific value 32, 33
    imports 16, 18, 19, 35
    production 16
        home 25, 32, 35, 39, 233
        non-farm 57
        in wartime 18
Food (Defence Plans) Department 18
Food Production Department 26
Forestry Commission 22, 168, 169
forms 113, *115*
    layout *4, 6, 7, 8*, 114
fruit 17

Gardiner, H. Rolf 17, 18
Gee, Evelyn 121
gender bias 197
General Views 21
geo-referencing 158
Germany 33
GIS (Geographic Information System)
    158, 159
grassland 17, 27, 35, 59
    fields ploughed up 131–132
    survey (Stapledon and Davies) 3, 32
Grassland Improvement Station 34
grazing
    rights 122, 177, 179–180
    rough 59, 75, 134, 201, 202
guaranteed minimum prices 15, 18, 39

Hall, A.D. 39
Hampshire 38, 179, 198, 224, 234
Hayter Hames, G.C. 230
Henderson, George 31, 42
Herefordshire 10
    engines 185
    landed estates 165, 167
    maps 144, 145, 147, 148, 149, 150,
        152
        fragmented holdings 155
        presentation 157
Hitler, Adolf 33
Hitler's Whistle 98
holdings
    name of 118
    plough-up 214–218
    size 35, 175–176, *175, 176*, 193–194
    type 62

amalgamated 101–102, 117
    multiple 68, 116
Hollerith machines 82, 83, *84*, 85–86, 87,
    197
Home Guard 205
Hops Marketing Board 16
horses 135, 180, 181, 183
Horticultural Return 5, 7, 113, 118,
    135–137, *135*
hospitals 171
Hudson, Robert S. 43
Huntingdonshire 153, 154, 156, 169
Hurd, Anthony 234

imports
    duties 16
    food 16, 34
    oil 180
    regulation 18
India 16
industrial concerns 170
insect pests 124
*Instructions for the Completion of the
    Primary Record* 56, 66, 68, 73
Integrated Admission and Control System
    (IACS) 233
interwar agricultural surveys 21–25
interwar policy developments 15–19
Isle of Man 110
Isles of Scilly 10, 108

4 June Returns *6*, 22, 113
    1941 *see* Census Return

Kent 45, 60, 83–84, 85–86, 99, 154, 179,
    198, 202
keys (maps) 75, 147–148
Kirk, J.H. 82, 88, 94

labour 134–135, 136
labourers 188
Ladybower Reservoir 147, 153
Lake District 152
Lancashire 42, 92, 156–157, 234
land, requisitioning 100, 202–205
land agents 163–164, 165
Land at War 234
Land Commissioners 20, 50–52, 57, 58,
    74, 98, 107, 108
land deterioration 16, 17
land girls *see* Women's Land Army
land nationalization 31, 39, 232

land ownership 161–162
   details of ownership 162
   discrepancies and errors 162–163
   landowners *see* landowners
   multiple ownership 163–164
   patterns of 164
      mapping 164–165
   tenure *see* tenure
land planning 54, 77, 78
land reclamation 182–183
Land Utilisation Survey 2, 22, 219
landed estates 82, 165, 166–167
landowners 57, 98
   institutions 168–171
   private 166–168
   women 194
Lend–Lease agreement 35
ley farming 17, 34
Liaison Officers 21, 30
libel 57, 98
Liberal Land Committee 1924/25 27
Lincolnshire 151, 156, 184
livestock *36*, 38, 134
Living Casualties 37
Lloyd George Survey 171, 237
local government 168
local protest groups 37
London 3, 215
Lymington, Viscount 17, 18

McDonald, Ramsay 16
machinery pools 181–182, *181*, 229
mailing lists 63
managerial classifications 26, 31, 56, 57,
      97–98, 236
   additional 58
   confidentiality 57, 93, 105
   differing standards 59
   Primary Return 126
   upgrading 102
   women farmers 191–193, *193*
manufacturers (tractors) 186, 187, 188
maps *9*, 102–103, *109*, **143–144**
   archiving 107–108
   authorship 151
   boundaries 5, 54, 77, 146
      colour coding 75, 146
   completion 74–78
   condition of sheet 145–146
   copies 150–151
   database variables 11
   dating 103, 149–150

   edition 144–145
   farm reference type 148, *149*
   fragmented holdings 154–155
   geo-referencing 158
   keys 75, 147–148
   marginalia 155–156
   military land use 68, 153–154, *153*,
      200
   miscellaneous information 156–157
   multiple holdings 68
   non-farming land 152–153
   number 144
   Ordnance Survey 102–103
      parcel numbers 67, 68, 75, **139**,
         148–149, *149*, 208, 210
      sheets *9*, 67–68, 75, 76, 119
   post-war use 77
   preservation 157
   problems 102–103
   progress of fieldwork 61
   scale 144
   shading 146–147
   stamps 151–152
   storage 76–77
   treatment of non-farming land
      152–153
   WAEC land use 154
market forces 16, 222
marketing schemes 16
matching process 60, 85, 86, 100–102,
      116
meat 16
mechanization 35, 180–184, *182*, 233
   fixed and portable engines 184–186
   tractors 42, *181*, 186–188, *188*
mice 124
Microsoft Access 2.0 11
Middlesex 151, 157
Midlands Sample 10–11, 113
   acreage figures 172–175
   Census Return 134
   crops sown *213*, 222, 223
   farm mechanization 184–188
   forms 114
   Horticultural Return 136–137
   land ownership 163, 167, 169, 170
   maps 144, 145, 146, 147, 148, 151,
      152
      fragmented holdings 155
      geo-referencing 158
      marginalia 156
   plough-up data *209*, *215*

Primary Return 116, 117, 121, 126, 129, 130
tenure *120*
military activities 35, 198
    damage by Home Forces 201
    enemy action 102–103, 206
    farmers and their families 205
    information
        sources 198–200, *199*
        types of 200–201
    requisitioning of property 100, 202–205
    training areas *204*
milk 16, 17, 23
Milk Marketing Board 16, 23
Milner Committee 25
Ministry of Agriculture 15, 19–21, 37
Ministry of Agriculture and Fisheries Act 1919 27
Ministry of Agriculture and Fisheries (MAF) 19, 28, 38, 39
    Economics Branch 93
    farm mechanization 183–184
    first Farm Survey 41–43
    National Farm Survey 1941–1943 *51*, 54
        circulars 56–60, 66, 67, 72, 96, 102
        finance 47, 88, 91
        offices 43
        personnel 43–44
        Statistical Branch (Lytham St Anne's) 5, 63, 64, 71, 73, 95, 96
        Water Supplies Branch 93
Ministry of Food 1, 21, 22, 37
mixed farming, remote 11
moles 124
Monthly Digest of Statistics 3
Morrison, W.S. 18, 21
motive power 137

National Agricultural Advisory Service 235
National Farm Survey (NFS) 1941–1943 *51*
    archival policies *see* archival policies
    documents *see* documents
    duration 3
    first Farm Survey 1940 41–43
    organization *see* organization of NFS
    problems *see* problems
    progress of *see* survey progress

purpose 54–55, 58
    survey process *see* survey process
National Farm Survey Records 3–10, *4*, *6*, *7*, *8*, *9*
National Farmers' Union (NFU) 17, 20, 37, 40, 93
National Milk Scheme 38
National and Regional Sample Parishes *12*
National Sample 11, 113, *115*
    acreage figures 171–175
    Census Return 133, 134
    crops sown *211*, *222*, *223*
    farm mechanization 184–188
    forms 114
    Horticultural Return 135–136
    land ownership 162, 163, 167, 169
    maps 144, 145, 146, 147, 148, 149, 150, 151, 152, 153, 154
        fragmented holdings 155
        geo-referencing 158
        marginalia 156
        plough-up data *209*, *215*
    Primary Return 116, 117, 121, 126, 129, 130–131
    tenure *120*
National Trust 169
New Forest 179
NFS of England and Wales (1941–1943): a Summary Report 93
non-farming land 152–153, 168
Norfolk 187, 221
Norman Conquest 229
Northamptonshire 156
Northbourne, Lord 17, 18
Northumberland 59, 150, 151
Nottinghamshire 10, 35, 193
    land ownership 167, 169, 170
    maps 147, 148, 150, 151, 152, 154
        fragmented holdings 155
        presentation 157

oats 15, 33, 34, 35, 222, 223
Ordnance Survey *see* maps
organization of NFS
    Advisory Economists and Advisory Centres 47–50, *48*, *49*, *51*
    County War Agricultural Executive Committees *see* County War Agricultural Executive Committees (CWAECs)

organization of NFS *continued*
   Crop Reporters 50
   Farm Survey Research
      Sub-Committee 52
   Farm Survey Supervisory Committee
      52
   Land Commissioners 50–52
   Ministry of Agriculture and Fisheries
      see Ministry of Agriculture and
      Fisheries (MAF)
   structure and information flows *51*
Orr, John 22
Orwin, C.S. 22, 39, 105, 232
Oxfordshire 22, 42, 156, 169, 219, *221*

Parish Lists 63, 71, 107
parishes *12*, 210
pasture 126, 127
   conversion to arable 32, 33, 34, 35
Peak District 10
Pearl Harbor 3
personal failings 98, 126, 192–193
pigs 16
plough-up campaign 15, 32, *36*, 131–132,
      206, *207*
   arable acreage targets 33, 34, 35
   crops sown 33, 222–223
   CWAECs 33, 35, 223–224
   farm mechanization 183
   information in the Primary Return
      crops sown 210–214, *211*, *212*,
         213
      field Ordnance Survey 208–210,
         208
      grass fields ploughed up 206, 208
      parish 210
      WAECs direction 214
   limits 34
   local and holding level 218–222,
      220, *221*, *222*
   local relations 37, 40
   national and regional extent 214–218
   payment per acre 33, 37
   quotas 35
   timing 33
Plymouth 103
policies, agriculture *see* agriculture
potatoes 16, 33, 35, 183, 223
price supports 15
prices 15–16, 22
   control of 24
   fixed 33
   guaranteed 15, 18, 39

Primary Record form (B496/E1) 69,
      70–73, 97
Primary Return *4*, 5, 52, 56, 113, 132
   acreage figures 114, 116, *174*
   arable land 126
   code no. 116–117
   completion of 68–73
   conditions of farm 123–125
   county 116
   crops sown 222–223
   district 117–118
   electricity 126
   Farm Survey form 64–66
   farmer 118–119
   fertilizers 127–128
   general comments 129–131, *129*
   grass fields ploughed up 131–132
   grazing rights 122
   management *see* managerial
      classifications
   military activities 198, 201
   name of holding 118
   Ordnance Survey sheet 119
   parish 118
   pasture 127
   plough-up campaign *see* plough–up
      campaign
   problems *see* problems
   progress of fieldwork 61
   record information 128
   tenure *see* tenure
   title 116
   water 125–126
problems
   finance 103–104
   with forms 101
   4 June 1941 Return 94–97
   maps 102–103, *109*
   matching process 100–102
   Primary Return 62–63, 97–100
Provincial Advisory Centres 5
Public Record Office (PRO) 10, 50,
      106–109, 110, 143, 164, 235
Public Relations Officer 21
publication 93–94
punched cards 82, 86

questions
   census returns 63–64, 66
   Primary Return 66
quotas 35, 233

rabbits 124, 201

railways 123, 170
rats 124
record keeping 38
Recorders 48–49, 69–70, 97
records
 access to 105–106
 Advisory Economists 110
 farm 110
 National Farm Survey 3–10, *4, 6, 7, 8, 9*
regional agricultural economists 24
regional decentralization 20–21
regional survey movement 22
reminder form 139
rent 138, 194, *195*
requisitioning of property 100, 202–205
research methodologies 10–13
reseeding 224
reservoirs 147, 153, **168**
Revised Instructions for Completion of Farm Records and Maps 57, 66, 68
roads 123
rooks 125
Rothamsted Experimental Station 2, 55, 87, 92
rough grazing *see* grazing
Royal Agricultural Society 2
rural industries 23
Rural Land Utilization 61, 105
rural society
 farms *see* farms
 land ownership *see* land ownership
 mechanization *see* mechanization
 military use of land *see* military activities
 plough-up campaign *see* plough-up campaign
 women and agriculture *see* women
rural surveys 40
Russia 3
Rutland 151

salaries 46, 50
sampling 3, 85, 86
 random (farms) 55, 83
 research methodology 10–11
 stratified 83, 93
Sandy Knoll Hotel 74
Scientific Food Committee 34
Second Front 206
Second World War 15, 219, 229

shading (maps) 146–147
sheep 201
Shropshire 70, 157
signatures (on forms) 70
silage 17
Sinclair, Sir John 1
slaughtering 17
soil 35, 123
 clay 180, 221
 fertility 33, 34, 35
Soke of Peterborough 157, 170
Somerset 157, 169
Somerset Farm Institute *207*
South Downs 10, 219
Southampton 102
Speed the Plough 39
398/SS (green form, census) *6*, 50, 64, 73, 74, 95, 96, 97
Staffordshire 35, 151
Stamp, Dudley 2, 76, 77, 82, 91, 105, 237
Stapledon, Sir George 17, 34, *207*
Statistical Account of Scotland 1–2
statistical analysis 83–87, 91, 93, 233
Statistical Methods for Research Workers 2
Statistical Society 2
stockpiling 32
storage 58, 108
Street, A.G. 16, 98
subsidies 16, 18
Suffolk 144
sugar beet 16, 23, 223
Summary Report of the Statistical Results of the War-Time Farm Survey of England and Wales 93, 218
Supplementary Form 5, *8*, 96, 114, 134, 137–139, *174*
 farm mechanization 183
 multiple ownership 164
Surrey 99, 154, 198
Survey of Buckinghamshire Farms and Estates 82
survey process 68–78
survey progress 60–63
 collation of data 55
 delays 58, 60, 61, 72
 main stages 52, 54
 summary, 1942 57–58, *61*
survey work 2–3, 26, 31
surveyors 46–47, 56–57, 69, 70, 98
 women 197

Sussex 10, 75, 99, *149*, 167, 184, 198, 235, 237
    Ashburnham 180
    Crown Estate 169
    military activities 198, 201
    plough-up 215, 219, *220*, 225
    women farmers 193, 194
Sussex Sample 10, 113
    acreage figures 171–175
    Census Return 133, 134
    crops sown *212*, 222, 223
    farm mechanization 184–188
    forms 114
    Horticultural Return 135
    land ownership 163, 167, 169
    maps 144, 145, 146, 147, 148, 151, 152, 154
        fragmented holdings 155
        geo-referencing 158
        marginalia 156
    plough-up data *209*, *215*
    Primary Return 116, 121, 126, 128
    tenure *120*

tabulation of records 82, 85, 103–104
tanks *199*
tenure 92, 165–166, 188
    length of occupation 137
    Primary Return 120, *120*
        details of ownership 162
        multiple ownership 161
        types of tenure 165–166, *166*
The Future of Farming 39
'the Great Betrayal' 16
The Land of Britain: its Use and Misuse 237
The New Anarchy 37
The New Morality 37
The Plough-up Policy and Ley Farming 34
The Way of the Land 34
tithe surveys 238
tractors 42, *181*, 186–188, *188*, 215, 216
travel allowances 48

University of Bristol 87
University College, Aberystwyth 22
upland farming 11

vegetables 17, 133
vehicles 185

WAEC 154, *181*
Walden, George 224
Wales 63, 184, *218*
    county
        Cardiganshire 150, 170
        Glamorgan 169, 194
        Merionethshire 144
        Monmouthshire 108, 144, 184, 194
        Montgomeryshire 150, 182–183, 225, 234
        Pembrokeshire 42, 167
        Radnorshire 151, 167, 216
        ploughed-up land 216, 218, *218*
War Cabinet 55
war policies 37–39
wartime administration 54
Warwickshire 34, 110, 154, 206
water 63, 91, 93–94, 125
water boards 168
water wheels 184
weeds 35, 124
Welsh Plant Breeding Station, Aberystwyth 34
Westmorland 59, 150, 157
wheat 15–16, 18, 35, 222, 223
Wheat Act 1932 16
Williams, Raymond 229
Williams, Tom 93, 238
Williamson, Henry 45, 230
Wiltshire 92, 234
women 189–190, *189*
    administrative role 196–197
    farmers 190–191, *192*, *195*
        grading 191–193, *193*
        size of farm 193–194
    labourers 194, 196
    landowners 194
Women's Land Army 28, 30, 42, *182*, *190*, 197, 224
wood pigeons 124
Woolton, Lord 3
Wye College 83, 233
Wyllie, J. 83, 91, 233

Yates, Dr F. 3, 55, 82, 84, 87, 92
50-year closure period 108
Yorkshire 216
    East Riding 151
    North Riding *46*, 60, 84, 85, 184
    West Riding 150, 170
Yorkshire Agricultural College 22
Young, Arthur 2